CW00967995

Hawthorn

Hawthorn

The Tree That Has Nourished, Healed, and Inspired Through the Ages

Bill Vaughn

Yale UNIVERSITY PRESS New Haven & London

Yale University Press books may be purchased in quantity for educational, business, or promotional use. For information, please e-mail sales.press@yale.edu (U.S. office) or sales@yaleup.co.uk (U.K. office).

Set in Bulmer type by Integrated Publishing Solutions, Grand Rapids, Michigan.
Printed in the United States of America.

Library of Congress Cataloging-in-Publication Data

Vaughn, Bill, 1948–
Hawthorn : the tree that has nourished, healed, and inspired through the ages / Bill Vaughn.
pages cm
Includes bibliographical references and index.
ISBN 978-0-300-20349-3 (alk. paper)
1. Hawthorns. 2. Trees—Montana. I. Title.
QK495.R78V38 2015
583'.73—dc23
2014043898

A catalogue record for this book is available from the British Library.

This paper meets the requirements of ANSI/NISO Z39.48–1992 (Permanence of Paper).

10 9 8 7 6 5 4 3 2 1

For Kitty

And every shepherd tells his tale
Under the hawthorn in the dale.

—MILTON

Contents

Preface

As a writer for magazines, I have traveled far and wide reporting about people, places, and events ranging from the trivial to the transcendent. Off the coast of Borneo I bribed fishermen to sneak me past the guards to an island where the first season of the CBS hit *Survivor* was being filmed. I walked across the bed of the North Sea at low tide to another island, this one off the coast of Holland, and back to the mainland before the tide returned, a peculiar and dangerous sport called mud walking that the Dutch seem to adore. I went around for a week with a busload of retired people taking in stage shows at Branson, Missouri. I spent a day in a machine shop in Austin, Texas, interviewing reality star Jesse James (before his famous divorce from Sandra Bullock). I bicycled around Provence on a gourmet tour and spent one night at the marquis de Sade's chateau and another at a small castle belonging to Diane de Poitiers, the accomplished courtier and mistress of Henri II. I built a wind-powered boat and sailed it both legally and illegally on railroad tracks in Montana. And I spooked myself visiting some of the eerie humming places the tribes of the Northern Plains consider sacred.

But even as I was wandering around the world, outside my own back door was the most compelling story I've come across so far. This is the drama of the hawthorn, a tree bound up with recorded human history for at least nine thousand years. After I stumbled across this extraordinary form of life and began studying it, I was astounded to discover that the role it played in the political history of Europe was indirectly responsible for my own existence. And so I decided that I needed to find out more about it, a serendipitous education that led me down a number of strange paths. The hawthorn's role as a potent symbol in Catholicism compelled me to try to understand the religion I had rejected as a motherless child. And the part it played in my own history drew me to the Irish countryside, where my family had struggled to survive, and to the Jesuit mission in Montana where my great-grandfather found his place in the New World.

While my personal connection with the hawthorn—our home, Dark Acres, is in the middle of a community of hawthorns stretching for hundreds of miles—drove me to write this book, I also believe that its political, religious, and natural histories are stories that need to be woven into a single narrative. These stories may come as a surprise to many readers, especially Americans, because we have largely forgotten the role the tree once played in North America among European newcomers and indigenous peoples alike.

At the core of my fascination with the hawthorn are its paradoxes. This is a tree that, armed to the teeth with pathogen-bearing spines, nonetheless harbors and nurtures a multitude of creatures. This is a tree that once served as an important icon in the spiritual life of pagans yet became an object of adoration in the new religion, Christianity, that tried to eclipse the old worship. And this is the tree, planted in hedgerows that made farming possible, that was used mercilessly to deny poor Europeans access to what once had been land shared by everyone. Now the tree offers patients with bad hearts new hope for a healthier life.

The story of the hawthorn is, in some ways, our own story.

Hawthorn

ONE

The World's Busiest Tree

There's no sense in getting killed by a plant.

—*DAY OF THE TRIFFIDS*

One morning during our first spring at Dark Acres I filled my chainsaw with fuel and oil and lugged it across a pasture to a tangle of twenty-foot trees in full white bloom. The sky was moody and overcast, glowing with that gauzy May light photographers love. Overhead a chevron of Canada geese passed so low I could hear the hiss of their wings. The air was perfumed with pennyroyal and the languid fragrance of cottonwood buds. A breeze blowing up the Clark Fork pushed hypnotic waves through the fresh green grass, lush after a week of warm rain, making our river valley look more like Ireland than the normally parched terrain of western Montana. Now that my wife, Kitty, and I were living in the country again, in the same sort of redneck backwater where I spent my motherless, feral boyhood, the only thing that could have made me happier on this perfect day was finding a hundred-dollar bill blowing in the wind.

On closer inspection I saw that these bushy trees were actually a single tree that had shot out eight trunks in all directions. It wasn't graceful like the weeping willow in our backyard, or majestic like the hundred-foot ponderosas in our forest. Blotched with crusts of blue lichen, the gnarled and twisted limbs arched down from its crown like the fingers of witches. Its gray bark was peeled and flaking. This looked like a tree only its mother could love.

The trunk causing our recent problem had grown parallel to the ground for fifteen feet and a yard above it, building a thorny wall of zigzaggy branches that embraced a confusion of vines and a length of web fencing that had been strung between pine posts, now rotted. Woven from steel wire into four-inch squares bound at the intersections with tight twists of a thinner gauge, the fence had been rusted and pitted by the weather, and warped and folded by the force of the growing tree.

I wondered why the ranch family whose cattle once wandered across this floodplain had used webbing instead of the odious barbed wire that snaked everywhere else through the forest. But recalling the pigs and their enclosures on my sister's cattle ranch in central Montana, I decided that pork-friendly fencing must indeed be the explanation. Whatever the reason, like the barbed wire, which I was beginning to replace with more horse-friendly post-and-rail fences made of pine, this nightmare union of briar and metal would have to go.

The day before, when I had gone out to bring in our quarter horses from the pasture, I was horrified to find Timer, our old brood mare, standing by the tree, head down, her left front hoof raised. Walking closer I could see that it was caught in the webbing. While concentrating on her work, which was the grazing she and the others were allowed four or five hours a day, Timer had somehow managed to step through a gap in the fence. I ordered Radish, our noisy red heeler, to back off. But he'd already launched himself into his daily chore of herding the horses back to their corrals, and couldn't be recalled. When Timer saw him charging, his yap now turned to full volume, she pulled back from the fence in alarm. I stopped yelling at him, expecting the worst.

Suddenly Timer was free. She trotted, limping, to a neutral corner away from the tree and threw a wild kick at Radish, who dodged it. I went to her, saying soothing things. When I picked up her foot I saw with relief that she wasn't cut, though in freeing herself she'd pulled off a shoe. The damage could have been far worse. Panicked horses caught up in wire sometimes slice their legs so deeply they have to be put to death.

I laid the chainsaw next to the tree and considered the problem. I figured if I could cut through the offending trunk, which was two feet around, I could drag it and its cargo of wire to our dump with a chain hooked to our pickup. I searched for a place that wasn't guarded by thorns so I'd have a clear shot with the saw, but the entire trunk was heavily armed. Using my pruning shears, I cut a pair of gnarly branches a half-inch in diameter from a superior branch that was twice as thick. When I tried to pull these cuttings loose, thorns bound them like Velcro. After considerably more snipping and tugging I had accomplished almost nothing. Losing my patience, I clipped the thorns from a section of the larger branch and pushed on it while tugging at one of the severed limbs. But my hand slipped from the embrace and the branch snapped back at my face.

Reflexively, I turned my head and stepped away. A stab of pain shot through my foot. When I lifted it a dead branch came along as well, one of its long thorns impaling me through the heel of my tennis shoe. The wound was already beginning to ache. Then I saw blood on my hand. Another spike had ripped a gash across the palm, neatly following the curve of my heart line.

Wincing in pain, I limped from the tree and sat down in the grass, for the moment defeated. A magpie landed in the maze of weedy branches atop an enormous, sloppy nest of piled twigs. When the bird looked down it cocked its head, laughed in that vulgar corvine way, and flew off in contempt. Sudden gusts of wind, cool against my skin and then warm, tossed the limbs of the tree around. I had the strong sense that the tree was celebrating, high-fiving itself, gloating about its triumph over the hapless human squatting on the ground. A shower of small, delicate white blossoms rained down on me. I picked a few off my shirt and raised them to my nose, expecting perfume. But they didn't smell sweet at all. They smelled putrid, with strong notes of sex. I realized then that the tree was fairly vibrating with swarms of something. But these weren't the bees and hummingbirds that mobbed the old flowering apple tree back at the house. These were flies.

What the hell *was* this thing? Some sort of mutant, like the berserk outer-space plants in that ludicrous science fiction movie I'd loved as a kid called *Day of the Triffids?* Or could it be some other kind of botanical carnivore whose seeds had escaped from a fiend's greenhouse and were blown here by the wind?

Well, whatever you are, enjoy yourself while you can, I thought, plotting revenge. *I will burn you out. I will poison you. I will loop a chain around your body and yank you from the ground with my pickup.* Well, probably not—this tree was such a fortress that trying to pull it out by the roots would only separate my truck from its bumper. I could feel my heart racing, my anger growing, and my blood pressure rising. My ears had begun to ring. When my hand finally stopped bleeding the cut began to burn. I wondered whether the tree had poisoned *me*.

The tenacious manner in which the tree had adhered to the fence made me think of the way ivy climbs the bricks around Wrigley Field, exploiting the brick walls for support and the heat they absorb. Maybe the tree was using the fence for extra protection. But with all its bristling thorns, needle sharp and indifferent to the pain of others, I couldn't imagine why the tree felt it needed more armor. How much security is enough?

The flowers were overtly erotic—shallow bowls a half-inch across made up of five white oval petals tinged with pink, arranged around ten reddish stamens. The leaves, the size of quarters and shaped like snowshoes, were shiny, serrated, veiny, and tough. Maybe a goat could find a meal here, but it didn't seem likely that anything else would consider the tree edible. Yet when I plucked a leaf and nibbled it I was surprised to discover a mild, earthy flavor reminiscent of a green tea such as oolong, except that it left a tang on the tip of my tongue. Spitting it out, I gagged. Yikes, what had I just put in my mouth? (I would later learn that young hawthorn leaves are used in salads, and during World War I across the British Isles they became a substitute for tea and tobacco.) The stiff purple thorns were an inch to almost two inches long.

As I began calming down, I realized that this tree was a fortress not just because of its adamant mass but also because other denizens of Dark Acres were using its protective cover and the merry kiss-off attitude it radiated as a refuge. Besides the magpies, a wild rose was thriving among the trunks, along with snowberries, a colony of wild raspberries, a stand of nettles, a chokecherry, some kind of red-barked bush, and at least two varieties of vines. From the deepest recesses of this briar patch came a rustling that made me wonder which animals were using this complex sanctuary to escape predators such as the bald eagles, foxes, bobcats, and coyotes that made their homes in these loops of river.

Waging war on trees wasn't what my wife and I had in mind when we bought Dark Acres. Kitty and I had fallen in love with these forests, beaches, and swampy sloughs, and we'd been appalled at the way some people on the river abused their land. One, a sullen mill worker from South Dakota, had even stripped away the brush, cut down the trees, and torched everything that wasn't grass, leaving acres of antiseptic lawn without a wildflower or a single hiding place for a grouse or a rabbit. Maybe he was homesick for the monotonous prairies of his native state.

But my run-in with the tree had humbled me. Although I knew a bit about the flora at Dark Acres because I grew up in a similar sort of jungle along the Missouri River, on the other side of the Rockies, I knew nothing about this formidable living being. My ignorance embarrassed me. *Well, we've been busy moving in,* I told myself, *trying to make the place safe for our horses, figuring out how to live in the country again, with its isolation and constant temptation to people who work at home to play all day instead.*

I lugged my chainsaw back to the shop. Then I grabbed a couple of unread field guides from my office, and headed back to the pasture. *Know your enemy,* I thought.

Yet after reading for an hour, I began to see the tree not as a monster or an adversary but as something more like a famous relative the family forgot to mention. A month later, after immersing myself in the vast lore of this tree, I felt as though I'd walked into my garden and discovered the world. Here was the crown of thorns and the burning bush. Here was the most famous tree in Britain, and the dearest tree in Christianity. Here was the fuel my Iron Age ancestors probably burned to forge the spears and swords that conquered Rome. Here was the hedge plant the English and their Irish allies used as a weapon against my great-grandfather. And here was the tree whose essence is used to treat people with cardiovascular problems all over the world.

I'm talking about the hawthorn.

Also known as thornapples, May-trees, and whitethorns, hawthorns belong to a genus of shrubs and small trees numbering as many as two hundred species native to the temperate regions of the northern hemisphere, the majority in North America. The scientific name for the genus, *Crataegus,* is thought to be drawn from the Greek word *krátys,* meaning "hard" or "strong," a reference to its incredibly dense wood. *Crataegus* is a member of the rose family of showy flowering plants, and its cousins include big-money food producers such as strawberries, apples, peaches, and pears. Although you can eat hawthorn fruit, and some species produce a yummy treat, most of these berry-like pomes are mealy and insipid, and contain cyanide-bearing seeds called pyrenes encased in hard shells. The fruit ranges in color from black to purple, red, and even yellow, and in diameter from a quarter-inch to an inch and a half. The name of the fruit, "haw," comes from the Middle English word for hedge. Along with its tough wood, the plant's thorns helped create impenetrable living fences that gave farmers thousands of years ago a way to protect their fields from roaming animals. (One species native to Montana, the fleshy hawthorn, *Crataegus succulenta,* brandishes vicious purple spikes more than two inches long. In what appears to be a botanical overachievement, these thorns sprout thornlets.)

Most hawthorns bloom in the spring, with flowers ranging from red to pink and white. When hawthorns grow wild in dense thickets, which they tend to do, the explosion of color and scent in May and the whirring of gazillions of pollinators around them make an impressive spectacle. The widespread groves in America and China, and the remaining hedgerows of Europe, offer vital habitats for many species of birds, insects, and beasts, and are a major player in their environment. Because of their floral extravagance in the spring, followed by the dark, saturated green of their summer foliage and the colorful ripening of their bountiful fruit in the fall, hawthorns are widely planted in gardens and yards.

Of all the beings that live at Dark Acres the hawthorn is the one that I've come to think of as the king—and the queen (its flowers are hermaphroditic). This is not just because we have more hawthorns here than any other species of tree. But while the cottonwoods and ponderosas are bigger, the foxes more photogenic, and the great blue herons and ospreys more theatrical, it's the hawthorn that's intertwined with the politics, religion and literature of Europe and the early rise of agriculture in America, as well as thousands of years of Chinese culture.

While exploiting humanity to spread itself around the globe, the hawthorn has not only been a source of wealth and inspiration; it has also been the cause of much misery. Stock-proof hawthorn hedges in Germany and Britain have been used for centuries to mark the boundaries of property and, like barbed wire in Montana, to keep cattle and sheep from wandering. They were also used by the landed gentry in Ireland and Britain to wall out peasants from what had been land whose bounty was used in common by people of all classes. Yet owing to the tree's mystical associations, it's considered such bad luck in Ireland to harm a lone hawthorn that roads are routinely routed around them. Some cultures regarded the hawthorn as supernatural, and others believed—and still believe—that it is invested with curative and magical powers. During Christianity's march to dominance in the Western world, some elements of the church tried to link the hawthorn with witchcraft as part of its campaign to suppress pagan beliefs and local languages by imposing its own iconography on native symbols and banning, for example, Gaelic in favor of the English spoken by Ireland's overlords. In Germany the hawthorn was associated with death and used in funeral pyres to create the divine smoke that would loft the soul skyward. In Great Britain the tree is at the core of the New Age revival of the old Celtic festival marking the rebirth of life. During the era of sailing ships many vessels bore the name *Mayflower.* Along with whitethorn, maythorn, May, and quickthorn, "mayflower" is another nickname for a species of hawthorn botanists call *Crataegus monogyna.* We don't know what the ship that brought the first English settlers to America looked like, but carved on the stern of a replica that sailed from England to Massachusetts in 1956 was a hawthorn flower.[1]

Copious references to the hawthorn reveal it as an enduring literary symbol. It figures dramatically in the pagan folklore of Europe, medieval interpretations of the Old Testament and the drama of the Crucifixion, and the stories of the Ojibwa Indians. In Sylvia Plath's scary, dreamlike poem "The Bee Meeting," the speaker's neighbors lead her to a circle of hives in a grove of blossoming hawthorns. They're dressed in protective black clothing; she's wearing a summer dress, isolated and vulnerable. "Is it the hawthorn that smells so sick?" she asks. "The barren body of hawthorn, etherizing its children."[2] This multi-layered poem is open to a variety of interpretations, but what doesn't need to be analyzed is the paradox of the white-blooming tree's evocation of death.

In the fantasies of J. K. Rowling, wizards don't choose wands, wands choose wizards. Harry Potter's first wand is made of holly, with the tail feather of a phoenix embedded in it. After he wrests away Draco Malfoy's wand, it shifts its allegiance to Harry, a crucial factor in Harry's final victory over Voldemort. Draco's wand is made of hawthorn wood with a strand of unicorn hair embedded in it. According to Rowling, a wand made of hawthorn is good for the magic that heals, as well as being adept at putting curses on people. But users have to be careful with it. In the hands of a dilettante its spells can backfire.[3]

Hawthorns also figure in Marcel Proust's seven-volume voyage into his past, *In Search of Lost Time.* The narrator tells us that he first "fell in love" with hawthorn blossoms during month of Mary church services (hawthorns blooming in May have been associated with the Virgin Mary for centuries). "The hawthorn was not merely in the church," Proust writes,

> for there, holy ground as it was, we had all of us a right of entry; but, arranged upon the altar itself, inseparable from the mysteries in whose celebration it was playing a part, it thrust in among the tapers and the sacred vessels its rows of branches, tied to one another horizontally in a stiff, festal scheme of decoration; and they were made more lovely still by the scalloped outline of the dark leaves, over which were scattered in profusion, as over a bridal train, little clusters of buds of a dazzling whiteness. Though I dared not look at them save through my fingers, I could feel that the formal scheme was composed of living things, and that it was Nature herself who, by trimming the shape of the foliage, and by adding the crowning ornament of those snowy buds, had made the decorations worthy of what was at once a public rejoicing and a solemn mystery. Higher up on the altar, a flower had opened here and there with a careless grace, holding so unconcernedly, like a final, almost vaporous bedizening, its bunch of stamens, slender as gossamer, which clouded the flower itself in a white mist, that in following these with my eyes, in trying to imitate, somewhere inside myself, the action of their blossoming, I imagined it as a swift and thoughtless movement of the head with an enticing glance from her contracted pupils, by a young girl in white, careless and alive."[4]

All these forays into the natural and cultural history of the hawthorn opened up a fascinating world of associations, but the discoveries this investigation led to also revealed something more personally compelling: my family history. Thanks to the hawthorn, I was introduced to my great-grandfather—not the man, since he died thirty years before I was born, but the story of the man, an illiterate Irish Catholic peasant who was forced to emigrate by the Potato Famine and the unbearable economic hardships created by the enclosure of common land.

Now that I know more about the hawthorn I'm glad I lost my battle with it. The hawthorn is responsible, at least in part, for the accidents of history that brought me into the world. And in a transcendental symmetry most people rarely experience, I learned that bound up in its fruit may be a safe and effective treatment for the high blood pressure that is a chronic genetic flaw of my family on my father's side.

In late September, after our discovery that Dark Acres was a preserve of wild hawthorns and the plant and animal life they harbor, Kitty and I were riding Timer and a gelding named Jack Reed through our forest when we came across a curious pile of animal shit in the middle of a trail. I dismounted and handed Kitty the reins. Poking at it with a stick I thought at first it might have been left by a raccoon, but there was too much of it, with no sign of the little pink crayfish from the river that these cranky, invasive beasts love to eat. Instead, it was chock-full of the seeds and skins of purple hawthorn fruit, plus chokecherries and rosehips.

And then we saw its probable source, sitting on his haunches not far away: a black bear, curious and brazen. The horses, which are usually hysterical around bears, seemed unconcerned, which made me guess that this particular bruin had been hanging out at Dark Acres long enough for them to become used to his smell. And they had probably seen him before, maybe every day, as he wandered around to graze, just as they were doing. With so much fruit, no bear fattening himself for his winter den in the stony palisades of the Bitterroot Mountains across the river would have the slightest interest in attacking a dangerous horse ten times his weight. Radish barked furiously, a mohawk of fur rising along his spine. But when the bear made no move to run away the dog simply glared, trying to stare him down.

I got back on Timer. As we headed to the house I turned around and saw that the bear was following us. When I looked again he suddenly evaporated into a wall of hawthorns so dense at its core that no sunlight reached the ground.

TWO

Under the Hawthorn Tree

> Could you but see the strange-looking characters . . . carelessly trudging along with a pipe in the corner of their mouths; the big shilelahs sloped across their shoulders, and all their worldly possessions tied up in a dirty bundle, suspended at the end of it—except, perhaps some half-dozen brats, one of whom rides *pick-aback,* while the mother brings up the rear with the other five—you would indeed be astonished how they drag out what must appear to everyone but themselves so miserable an existence.
>
> —JOHN BARROW

In Ireland it's not unusual to come upon a strange, solitary old tree swaying in a field, the tracks of a tractor swerving around it in the dirt. That the farmer, his father, and probably his grandfather took such obvious care to spare this tree says something about these farmers. It also says something about the tree. A rowan or a willow unlucky enough to take root in a valuable tract of barley or oats would soon go under the plow. But this lone sentinel is a hawthorn, more of an icon than a plant, holding an intensely symbolic and emotionally charged place in the cultural, religious, and political history of the island.

Pilgrims to this day tie ribbons and bits of their clothing to certain hawthorns, called "rag trees" or "clootie trees," as a supplication for health, money, or love. Flowering branches of hawthorn were hung on doors to keep out faeries, and put up in barns to encourage cows to produce more milk. The purity of its white blossoms was a central symbol in the month of Mary observances after the cult of the Virgin spread across Europe to Ireland during the Renaissance. But as one of the paradoxes in the body of folklore associated with the tree, it was believed that bringing a flowering bough into the home would invite bad luck or could even be a harbinger of death.[1]

The hawthorn was still considered a potent source of supernatural power when my great-grandfather was baptized on 1 March 1838 in a rural backwater of County Waterford, which lies on the southeast end of the island. In the Irish Catholic tradition of the era, that date was also recorded as his birthday. Thomas was the fourth child of seven or possibly eight born to Edmond Moran and Honora Bridgit Barton Moran, illiterate peasants who worked as farm laborers in the Roman Catholic parish of Mothel and Rathgormack below the Comeragh Mountains.[2] Encouraged by the church, which prohibited contraception and abortion, the fecundity of people such as my great-great grandparents was responsible for the 400 percent rise in Ireland's population between 1700 and the entry of Thomas into the world. Like most of the poor in rural Ireland, the Morans worked a small plot, growing potatoes and maybe a few turnips. They probably built their own shelter on their employer's land or moved into a hovel built by a similar family before them. This would have been a squalid, wattle-and-mud cabin without windows or a chimney, the roof a few poles covered with sod, barely high enough to allow the family to stand upright, the floor a hollowed-out ditch.[3]

The nearest town of any substance was Carrick-on-Suir a few miles north, which lay on the county's major thoroughfare, the River Suir (pronounced "sure"). Founded during the thirteenth century, by 1800 the town was bustling with manufacturing businesses ranging from wool cloth to candles. Much of its prosperity was made possible because the countryside, including the Morans' parish, was fertile and verdant. But within two decades of my great-grandfather's birth, Carrick's economy had collapsed because of falling prices for agricultural commodities and the fact that the hand-woven wool produced in the homes of townspeople could not compete with the machine-woven textiles of England.[4] In 1834 an English gentlemen named Henry Inglis who was visiting Carrick noted that he was "struck with its deserted falling off appearance, with the number of houses and shops shut up and the windows broken, and with the very poor ragged population that lingered about the streets. . . . I found the price of labour here lower than I have found it anywhere. . . . Many hundreds of unemployed labourers could have been got by

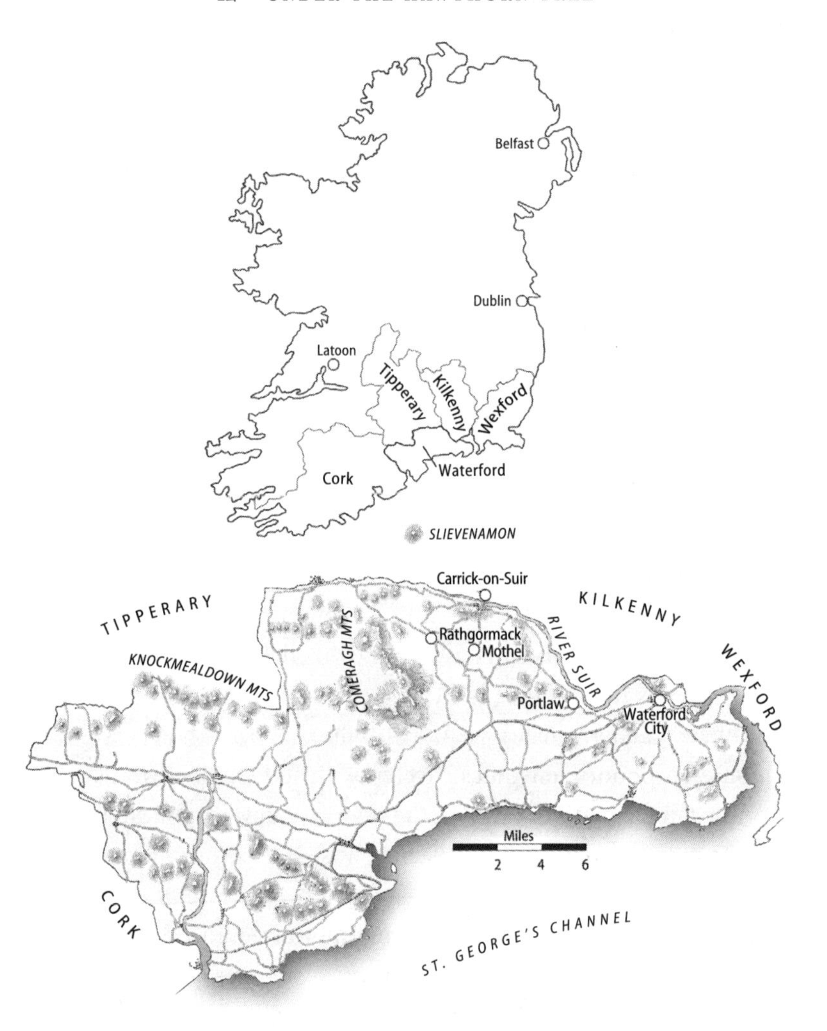

Ireland, showing the southeastern counties with (*below*) County Waterford in 1837 (map by the author)

holding up one's finger."[5] Seventeen miles downstream from Carrick was the city of Waterford. Founded by Vikings in 852, it's the oldest city in Ireland. Although by Montana standards the twelve peaks of the Comeraghs are more hills than mountains, the landscape, largely treeless and ground down by glaciers, is dramatic. Separated by extensive hedges of hawthorn, blackthorn, and holly, the fields are soaked by an average of thirty-two inches of rain a year and turn such an intense green in the spring that they seem to shimmer. Although County Waterford is located at the same icy latitude as Calgary, Alberta, the air is moderated by the Celtic Sea, which rarely allows winter temperatures to drop below freezing.

But on the day Thomas was born, this gorgeous countryside was the home of a beleaguered people. The Morans, like most of the Irish underclass, eked out a precarious existence one accident away from catastrophe. In the spring they planted the sprouted eyes of the potatoes saved from harvest the previous autumn into a "praty garden," or potato patch, fertilized from a mound of manure piled near the front door so the family could guard it. The shallow rows where they laid these seedlings could be dug without complicated farm implements or even much labor. Sometimes the eyes were simply dropped in a line and covered with slabs of peat in what were called "lazy rows." The process of planting, weeding, fertilizing, and harvesting the potato was much less demanding than that required for other agricultural commodities. But the result was a nearly perfect food.[6] Although monotonous, a steady diet of potatoes supplies twice as much daily protein and essential elements such as iron and calcium than was required by the average Irish adult, the scrawny 140-pound men and 90-pound women of the day. The potato is also low in bad fats and high in ascorbic acid, which prevents scurvy. And it was well-suited to the oppressive system of land ownership, in which peasants did not have access to enough land to grow grain or raise sheep or cattle. It didn't matter that potatoes don't keep well, like wheat and oats, because peasants such as the Morans didn't have the resources to build their own silos or barns for storage. It's no wonder the English rulers of Ireland and their Irish subordinates considered the potato "the lazy crop," the source of Gaelic indolence, which in their minds was tied up with an addiction to the pope.

Like other farmworkers, the Morans were probably not paid by their employer in coin but rather with food, a garden, and a roof (however shabby) over their heads. In the British Parliament's 1837 Report on the Poor of Ireland, the Morans would have counted among the 2,385,000 people who suffered in the summer months, before the potatoes were harvested, from malnutrition and the diseases it invites. This appalling figure represents almost a third of the island's people. Like most of the denizens of County Waterford included in the 1841 census, the Morans would have been considered a Class III family, people "without capital, in either money, land or acquired knowledge."[7]

One August afternoon I rode south from Carrick on my bike to take a look at Moran country. The narrow two-lane road rose from the valley of the Suir toward the Comeragh Mountains. I stopped at the churchyard in Rathgormack and wandered among the headstones looking for my relatives. But none were to be found. Nor were any noted at the old cemetery nearby, perhaps because what headstones survived had been so thoroughly weathered they were mostly illegible. I stopped to watch a farmer and his family herd their cattle on foot down the pavement from one of their pastures to another, both fields enclosed with hedges made of hawthorn and blackthorn. The red and purple fruit was abundant, and the trees were full of chattering birds. The manicured fields that spread out toward the mountains were so lush and picturesque they bordered on the surreal. It was hard to believe that this had been the site of so much misery only three generations ago. And it took a leap of imagination to see the wretched hovels that would have lined the empty rural lanes that in 1841 were the arteries of a countryside that was home to far more people than it is today.

Thomas Moran's first memories would have included a typical family dinner because the meal was the high point and focus of the day. In the smoky light of a fire fueled by peat quarried in nearby bogs and then dried in the sun, he would have sat with the others in a circle on the floor or outside around a basket of hot potatoes. If they were lucky that day they might have come into possession of some buttermilk, a small bowl of which would have been poured for each person as a dip for the potatoes. There might also be another baked cellar vegetable, maybe a turnip (at Halloween turnips were hollowed out to make jack o' lanterns; pumpkins replaced them when the Irish fled to America). In the winter, when night fell early, the Morans would go to bed soon after eating, on crude beds consisting of straw ticks and a few rags. And like most Europeans of the time, they would sleep in two stages; first waking after midnight, when the children would play while the adults fussed with household chores, then going back to sleep until dawn.

Like most peasants, children and adults alike, the Morans likely dressed in rags. Since they could not afford shoes they went barefoot in all seasons. During the winter Thomas wore a heavy cotton pinafore with deep tucks running laterally across the skirt so the dress could be lengthened as he grew. When he reached puberty he probably inherited a pair of hand-me-down leggings from one of his older brothers. Dressing boys in skirts was a common practice among Irish peasants, partly because skirts were easier to sew than britches. However, since his clothes were not waterproof he would be compelled to hole up in the cabin or under a tree when it rained, which happened often.

In the summer Thomas might bathe in the stream where the family's tattered clothes were occasionally washed, but in the winter a warm bath was a rare event requiring the laborious heating of water over a peat fire. Consequently, he was probably plagued by head lice, even though his parents trimmed his hair with a knife. It is unlikely that he had any real toys; he would have made his own fun with mud and sticks and stones. Like kids today he would have climbed trees, floated hunks of wood in streams, and played tag and hide-and-seek. He might have fished, carrying the spike of a hawthorn in a pouch for good luck, and tried to snare rabbits. His religious instruction would have been limited to the Bible fables his parents recited from memory and the occasional mass the Morans might attend in the village of Mothel or Rathgormack. Like most of the rural poor, he spoke Gaelic. It is not known when he began to speak English, the language of commerce and landlords and the law. As for formal education the Morans lacked the money to send him to a private school, and the free National School that was founded in 1840 in Coolnahorna was at least a four-mile walk away. There is no evidence he even attended one of the clandestine "hedge schools," which were intended to give Catholic children a rudimentary education (these will be discussed in Chapter 3, below). As late as 1870, when he was thirty-two, his letters were written for him by other people. So Thomas was fated for a life in the "brown industry"—farm labor—in an impoverished parish where uneducated, superstitious people were still afraid of mirrors.

As soon as he could form a sentence in his ancient native tongue Thomas would have been immersed in the vast folklore of Ireland, an oral library of stories, myths, and superstitions passed down and re-imagined by a hundred generations. Some of these originated with the

Celts, who spread across Europe and into Ireland beginning 2,700 years ago, absorbing the animistic beliefs of farming cultures that had grown up on the island since the end of the Ice Age. Because the Celts left no written records, what little is known about their culture was recorded by Greek and Roman observers. The Roman philosopher Pliny the Elder described an educated class of Celts called Druids, who venerated oaks and the parasitic mistletoe that grows on them, gathering in groves to perform their rites. In Ireland, oak worship was broadened to include a trinity of sacred trees: the oak, the ash, and the hawthorn. This may have come about after landowners began planting hawthorn hedges around groves of the other trees, which supplied valuable timber, in order to protect them from grazing cattle and lightning, which hawthorns were thought to repel. By the arrival of the nineteenth century, respect for hawthorns in Ireland was tied up with the belief that they harbored faeries, who lived among humans but were rarely seen.[8] At night the Fair Folk came out to dance and make love in the moonlight under the gnarled branches. If you wanted to see one of these bothersome runts that's where you'd go.

Why anyone would want to, however, says something deeper about the character of the Irish. The creatures of the *sidhe* were not adorable and effervescent like Tinker Bell in the Disney version of *Peter Pan,* nor were they regal and eloquent like King Oberon and Queen Titania in Shakespeare's *A Midsummer's Night Dream,* whose court met under the hawthorns. At their best Irish faeries were pests. At their worst they were horrid, deformed, and frightful goblins that wandered the countryside causing death and destruction. Occasionally, and at great personal risk to the supplicant, a faerie could be exploited to find out something about the future.

It was believed that the offspring of faeries often died before they were born, and those that survived birth were usually stunted or deformed. Faerie parents might then steal into a cottage and exchange their defective child for a human baby. These faerie babies were known as changelings, and they were usually identified by temperament rather than appearance: a child who constantly howled and screeched except when calamity crashed down on the household was probably a changeling. Soon after its arrival a changeling would grow a full set of teeth, its legs would wither, and its hands, covered with a light, downy hair, would become as gnarled as the talons of birds.[9] It was thought that children who were not baptized or were overly praised invited the scrutiny of faeries and ran a special risk of being stolen. A crucifix or iron tongs placed across the cradle was considered an effective talisman, as was an article of the father's clothing laid over the sleeping child. Because hawthorn thickets were the dwelling places of faeries, wreaths made from the branches of the tree to this day are put on doors to appease these bothersome and malicious forces.

The Morans apparently took no chance of losing Thomas. Soon after he was born he was baptized by John Condon, the parish priest of Mothel and Rathgormack, and the baptism was promptly recorded in the ledger. The heavy flannel dress he likely wore until puberty, in addition to its other advantages, was intended as protection, to make him look like a girl, because faeries had little interest in stealing females. Believing in both the Christian and the pagan spirit worlds did not pose a contradiction for the Morans. When appealing to unseen forces, poor people have always hedged their bets.

Thomas may have grown up in an impoverished material world, but his mind teemed with a phantasmagoria of faeries, witches, ghosts, elves, sprites, the seductive mermaids with webbed hands and feet called merrows, the banshees that howled when someone was about to die, and other supernatural beings that stalked the forests and ruled the night. Legions of Christian spooks—angels, demons, the devil, the Holy Ghost, apparitions of the Virgin—also kept him company. It's a wonder he ever got any sleep. But his whole world changed in 1845, during his seventh year, when the greatest tragedy in Irish history struck.

It began quietly. During that rainy summer a fungus-like pathogen from Mexico called *Phytophthora infestans,* or "late blight," made its way to Ireland by means of bat guano, imported from Peru as fertilizer, and attacked the potato crop. After 1800 potatoes had been the principal crop of the parish, supplanting oats, apples, and milk as the primary source of nutrition. For peasants such as the Morans it was almost the only food they had. Potatoes rotted in the ground, turning into putrid black bags that emitted a nauseating reek. As the blight ran its course over the next seven years more than a million people died of starvation and diseases such as dysentery, typhus, and fever caused by malnutrition. (Late blight is still such a problem in the 4 percent of Ireland's arable land planted with potatoes that some agronomists are calling for an end to the government's ban on genetically altered crops.)

The devastation suffered by Irish tenants and farmworkers was at best the result of depraved indifference on the part of the landowners, at worst a symptom of naked class warfare. The island's English overlords rationalized the tragedy as proof that the theories concocted by Thomas Malthus about the dire consequences of unbridled population growth were becoming a reality. "The land in Ireland is infinitely more peopled than anywhere else," the Reverend Dr. Malthus wrote in 1817, "and to give full effect to the natural resources of the country, a great part of the population should be swept from the soil into large manufacturing and commercial towns." A more coldhearted view was expressed in an 1847 letter from an Irish politician named Lord Ormonde to Lord Bessborough, Queen Victoria's head of the Irish executive. Ireland, he said, had two million more people than it needed.[10]

The eyewitness accounts of emaciated toddlers pleading for food and bloated corpses sprawled beside the road have been central to Ireland's sad modern narrative, inflaming passions on the island for 150 years. Starving peasants from the Morans' parish wandered from farm to farm begging for food. One week two men who had starved to death were found on the road from Carrick to Mothel; they probably died there instead of inside their shacks because they had been evicted. People were lucky to get anything to eat even once a day. In the spring and early summer they foraged for food in the hedgerows, dining on hawthorn leaves, dandelions, and vetch. In autumn they picked wild blackberries, haws, and the purple sloes of the blackthorn bush, a hawthorn cousin in the rose family. The buds of nettles were used to make soup. Beggars might stop at a farm and ask for "thrawneens"—the stems of hay remaining after the seeds were removed for sale. In 1846 the tariff on imported corn was removed so poor people could buy cheap, dried maize from America. But unless it was ground into meal, a milling process that was rare in Ireland, maize was indigestible. It came to be derided as "Peel's brimstone," after the prime minister of Britain, Sir Robert Peel, the architect of the scheme. None of these emergency foods could compare with the potato in taste and nutritional value. Haws from the one-seed hawthorn, *C. monogyna,* common in Irish hedgerows, although nutritious, are pulpy and practically flavorless (they have a slightly tart aftertaste). But in the autumn, when they ripened, haws were sometimes the only food people could find. Paradoxically, it was believed that after a season when hawthorn fruit was bountiful, scarcity would surely follow.[11]

Had they lived in Carrick or Portlaw, the Morans might have been given soup tickets by generous townspeople who could afford to be charitable. Or they might have joined the various robbers who took food by force. Tales of these crimes were rampant. Boats on the Suir carrying grain were pulled over at gunpoint and looted. Highway thieves stole butter. Sixty-nine fattened pigs being herded to Carrick for shipment to Waterford were dispersed by thieves, who stole every one.[12]

Catholic bishops in the province of Munster, which includes Waterford and five other southeast counties, decreed with an absurd sense of the obvious that poor people could, on fast and abstinence days, eat whatever they could find. The priest who baptized Thomas Moran, John Condon, signed the minutes of a committee meeting in 1846 reporting the "great destitution and distress" that plagued the area. That year farmers in the parish planted only about seventeen hundred acres of potatoes, a little more than half the amount planted the previous year, sowing oats, barley, and turnips instead. But that potato crop failed, as well.[13] Meanwhile, other commodities such as corn, barley, and dairy products produced in Ireland were being exported to feed the English while the Irish starved simply because prices were higher in England. The politicians at Westminster, who were advocates of free trade, believed that the Irish had brought their troubles on themselves because they had not embraced capitalism.

A best-selling children's novel written in 1990 and a subsequent television adaptation brought home to a contemporary audience the emotional devastation caused by the Great Hunger and its toll on family life. Set in Duneen, a village on the coast of the Celtic Sea in Cork, the county adjacent to Waterford, *Under the Hawthorn* depicts the plight of the O'Driscoll family, who after two years of lost crops are forced to seek food and work elsewhere, and are split up in the process.[14]

Thomas Moran, his two sisters, and three of his brothers somehow lived through the Great Hunger. But another brother, James, did not. He probably died of "famine fever" brought on by malnutrition—typhus, spotted fever, or relapsing fever, which is spread by lice and ticks. But it was not just the failure of the potato crop that either killed or forced the emigration of more than one-quarter of Ireland's 1845 population of more than eight million, 80 percent of whom were Catholics. By 1870 half of Ireland's acreage was controlled by only 750 families, some of whom were absentee landholders living in England. In a brutal system of land ownership called conacre (a corruption of "corn-acre"), these landlords rented arable fields to tenant farmers, who in turn sub-rented strips and plots as small as an eighth of an acre, called "sniggers" (among other names), for a period of no more than eleven months at a time to ensure that landless peasants could not establish a lasting relationship with the farmer or the land. Because of Ireland's enormous population growth before the famine, demand for land was high, and landlords could charge anything they wanted. They shielded themselves from their subtenants by using middlemen to collect the rent. Landlords were taxed for the cost of relieving the misery of the destitute on their property by order of an 1838 law, and might choose to reduce their tax burden by evicting their renters. Ten percent of all farm production was taken by the Protestant Church of Ireland. Catholics obviously got no religious benefits, or any other kind, from their coerced payment of this "tithe."[15]

Compounding the misery was the landlords' inhumane practice of denying peasants access to land where they had once grown crops, gathered firewood, cut peat, and foraged for food. This was accomplished by planting fast-growing, impenetrable hedges of hawthorn, blackthorn, and holly around fields to keep sheep and cattle in, and people out. In Britain, the government sided with landlords against the landless by passing more than five thousand Enclosure Acts after 1760 that privatized almost ten thousand square miles of common land that for centuries had provided for rich and poor alike. Although nurserymen in England had been making a living supplying hedge plants since the fourteenth century, the two hundred thousand miles of hedgerows planted between 1750 and 1850 required at least a billion plants, a big business generating fortunes for nursery firms and earning good livings for men who knew how to plant and prune hawthorns in a way that made a row of them impregnable (a process that will be described in Chapter 4). Although some of these commons had been wild, most were cultivated. In Britain enclosures represented almost a third of the arable land. The confiscation constituted a dramatic break with the open-field system of agriculture that had spread across Europe during the Middle Ages, when the king awarded a lord of the manor rights to two or three large pieces of land several hundred acres in size, and the lord required his peasants to farm it for him. In return, the peasants got the right to work their own fields and pasture their own animals. Included in a manor's landholdings were forests and pastures that were also used in common.[16] Some landless peasants tried to make a go of it on "waste" land—for example, bogs or rocky ground.

The enclosing of small parcels with hedges had occurred sporadically since the twelfth century. These tended to be actions that were taken by consensus, and they could be halted if any of the parties objected. That slow process changed dramatically with the Industrial Revolution and the rise of capitalism, when land came to be regarded as private property that could be bought and sold. Ownership became concentrated in fewer and fewer hands as peasants were driven from the countryside to work at factories in the cities. Britain's wars with the French and with the American colonies, and then its wars against Napoleon, created food shortages that Parliament and the crown decided could be dealt with only by increasing the productivity of agriculture (this thinking would occur again after World War II, with an opposite outcome for England's hedgerows). Peasants were considered inefficient farmers because they produced only enough food and fiber to pay the rent and feed and clothe themselves. Also reducing productivity was the fact that one of a manor's fields had to be left fallow every year so it could recover its fertility. In addition, the open-field system had created an illogical hodge-podge of long strips of land called furlongs. The Enclosure Act of 1845 streamlined the process of seizure by creating Enclosure Commissioners, who could award landlords additional property without going to Parliament for permission. In Ireland during the eighteenth century, greed and economic pressures similar to those in Britain resulted in many enclosures as well, especially in Munster Province. However, these were created in almost all cases not by acts of the Irish Parliament in Dublin but by the whims of landlords and the middlemen who made the critical decisions about their masters' farms. Most of these middlemen were Irish Catholics; some had secured leases that were in force for three lifetimes. Their interests were not in powerless tenant farmers and farm laborers but in extracting as much wealth from the land as possible.[17]

The dispossessed did not take these assaults lightly. Beginning in 1761 in County Limerick a secret organization called the Whiteboys began attacking middlemen and landlords they claimed were charging excessive amounts, called rackrents, to their subtenants, or enclosing land and evicting families to make way for beef cows and dairy cattle, following a huge increase in the demand for the commodities the animals supplied. Named for the white smocks these mostly teenaged Catholic farmworkers wore during their nighttime raids, Whiteboyism spread rapidly to Waterford and the other counties of Munster. The Whiteboys typically initiated hostilities by warning a middleman or a landlord of their grievances, thus giving the man a chance to mend his ways. If he ignored them they might kill his cattle, destroy his hedges, cut down his orchard, dig up his pastures in the effort to create more conacre and thus more potatoes, or torch his house, all of which earned them another name, Levellers, by the authorities. In Munster there were three outbreaks of Whiteboy agitation against enclosure and the tithe on the production of potatoes. But by 1786 these insurrections—a form of class struggle pitting poor Catholics against Catholic middlemen—petered out because of lack of organization and recurring food shortages that sapped the energy of even the most ardent insurrectionist.[18] Rebellion in Waterford flared again in 1848 as revolutions swept away monarchies all across Europe. But this time the leaders of the Irish seditionists were bourgeois gentlemen. Their goal was the repeal of the 1800 Act of Union, which had created a new nation, the United Kingdom of Great Britain and Ireland, stripped the Parliament in Dublin of its power to pass laws, and removed all barriers to the exportation of food and fiber from Ireland to England.

For the Morans, who were simply trying to survive from one day to the next, the jockeying for political power meant nothing. But to Young

Ireland, as the movement came to be called, Union was considered intellectually and morally odious. One of the group's most eloquent speakers was Thomas Francis Meagher (pronounced "Mar"), who was fifteen years older than my great-grandfather, Thomas Moran, and led a life that was the antithesis of his. Meagher was born into wealth in an enormous house on the quay running along the right bank of the Suir in Waterford. The building is now the elegant Granville Hotel, where I rode my bike one day to the establishment's Thomas Francis Meagher Bar.[19]

Meagher's father was one of a few extremely successful Catholic merchants who made their fortunes in Ireland's ports. In 1829, the year of Catholic Emancipation, which removed the ban on Catholics' holding office, he was elected Waterford's mayor. He sent his son Thomas to boarding school, and then to law school in Dublin, where the youth fell in with a group of writers, orators, and activists publishing a newspaper called *The Nation,* which agitated for repeal of the Act of Union. Meagher was a foppish dandy who loved books and fine clothes. He was also a gifted orator who inflamed crowds. He had no military training or experience with violence. But after the blight began to ravage the countryside, he began calling for an armed rebellion to end Great Britain's subjugation of Ireland.

Things came to a head after a huge rally attended by thousands was held in July 1848 on Slievenamon Mountain, a 2,300-foot rise in County Tipperary that looked down on the misery of the Morans' parish less than five miles away. Meagher rode up the mountain on horseback, wearing a tri-color sash and a gold-braided cap. The day after the rally he rode down to Carrick, where he met a contingent of supporters, and accompanied them back to Waterford. Revolution was in the air. Even the peasants, ground down by malnutrition, seemed ready to fight. While this ragged bunch possessed few firearms, all of them, including some women, were armed with whippy fighting sticks called shillelaghs made from blackthorn, *Prunus spinosa,* whose modern incarnations in Ireland are sometimes crafted for the tourist trade from hawthorn. These were the weapons of choice in the clan brawls ("faction fights") that were common sport at fairs in Carrick when Thomas Moran was trying to grow up.[20]

On 29 July a lethal confrontation broke out between a crowd of rebels and a detachment of police in the Tipperary town of Ballingarry, ten miles north of Carrick. It was over in an hour, leaving several insurrectionists dead. At the time, Meagher was in the Comeragh Mountains spending his nights lighting beacon fires, like most of the other leaders of the revolt, in order to give the appearance that the province was up in arms. He went into hiding when he learned about the events at Ballingarry. On 13 August he was arrested. Two months later he was found guilty of sedition and sentenced to death by hanging.[21]

One summer morning in a lukewarm mist I stood before a bronze statue of Meagher in Waterford that had been erected on a traffic island across the street from Reginald's Tower, a cylindrical stone fortress built by the Vikings in 1003. Leading the charge, his sword raised, Meagher is draped in the flag he designed for the Republic of Ireland.[22] Out of nowhere I experienced a vivid and jolting sense of déjà vu. I suddenly realized that I had seen a statue of Meagher somewhere else that looked very much like this one. When I remembered where, I realized that many years following the events of the failed 1848 uprising, after avoiding the death sentences pronounced on them both by the forces at the root of Ireland's relentless trouble, my great-grandfather, the illiterate peasant, and Meagher, the privileged nationalist, had crossed paths again—this time on the other side of the world.

THREE

The Celtic Forge

Should we succeed—oh! think of the joy, the ecstasy, the glory of this old Irish nation, which in that hour will grow young and strong again. Should we fail, the country will not be worth more than it is now. The sword of famine is less sparing than the bayonet of the soldier.

—THOMAS FRANCIS MEAGHER

Thomas Francis Meagher's theatrical ride up Mount Slievenamon at the climax of the 1848 uprising would not have been possible before 1782, when the law barring Irish Catholics from owning a horse worth more than five pounds was finally rescinded. The British Parliament decided that any mount worth less would be unsuitable for the sort of military shenanigans an insurrectionist might want to undertake. Until the end of the eighteenth century, Catholics were also not allowed to vote, become lawyers or judges, own firearms, serve in the military, inherit Protestant land, marry a Protestant, teach school, or adopt orphans. In the rare event that county authorities allowed a new Catholic church to be built, it had to be situated away from the main roads and constructed of wood, not stone. Not until the Roman Catholic Relief Act was passed by the British Parliament in 1829 were Catholics allowed to hold office.[1]

This onerous web of repression was forced on Ireland after James II, the deposed king of England, and his Irish Catholic allies were defeated at the Boyne River in 1690 by the Protestant army of James' son-in-law, William of Orange, who had been awarded the crown by Parliament two years earlier. In 1798 a new rebellion rose up, but it was ruthlessly put down. The 1829 "emancipation," following years of nonviolent agitation by the Catholic Association, brought some benefits to the middle class and better-educated members of Irish Catholic society. But for peasants such as the Morans life grew ever more difficult. In addition to sufferings brought on by the crumbling economy they were disenfranchised: women could not vote, no matter what their class, but neither could Edmond Moran, my great-great grandfather, because he didn't hold a lease worth more than ten pounds (about a thousand dollars in today's money). Until 1840 tenant farmers still had to pay taxes to the State of Ireland and a tithe to the Protestant Irish Church. After that year and until 1869, landlords were required to pay their tenants' taxes, which they extracted by raising rents and employing an enforcer to collect them. (One such enforcer, in neighboring County Cork, was my wife's great-grandfather, Patrick Dorsey Burke, a Catholic who worked for an absentee English landowner. One day in 1862 Burke finished making his collections and then sailed off to Boston with his boss' rent money.)[2]

Just as poor rural Catholics continued to practice their religion furtively in safe houses or under the open sky, a clandestine form of education sprang up in the 1700s to teach children reading, English, "maths," and in some cases the Irish bardic tradition of Latin, Greek, history, and home economics. These were the "hedge schools," so named because classes were often held outdoors, sometimes concealed behind a hawthorn hedge, with a child posted to serve as a lookout to warn the others if the law came snooping around. On cold days the children were asked to lug a brick or two of turf to the "school" so that the peat could be burned for warmth. ("I never carried the sod" became a common shorthand for lack of schooling.) Reading was taught using cheap chapbooks sold at country fairs that featured wild stories about freebooters and outlaws. When it came to the education of children, violations of the Penal Laws were largely ignored by local constabularies: during the prohibition against Catholic schooling, which lasted from 1723 to 1782, it appears that not a single hedge teacher was prosecuted. While children could attend Protestant schools sanctioned by England, most Catholics refused to enroll their children because these classrooms were seen as processing centers to force their assimilation. And most Catholics could not afford the schools. The hedge schools, which were free and unfunded, appealed to the poor and relied on the compassion and loyalty of educated adults, who might receive payment for their services in butter or other goods. In 1826 almost three-quarters of the "pupils" in Ireland attended hedge schools, where both Gaelic and English were spoken. The hedge schools endured until the end of the century, even after an 1831 law established a system of government-funded public National Schools.[3]

There are no records indicating that my great-grandfather attended a single class in any sort of school. In this sense he was like many Irish children in a country where in 1851 almost half the population could neither read nor write.[4] It is hard to gauge his level of ignorance as a child. Had anyone told him about the history of his parish or the origin of the name Moran? Had he ever seen a coin worth more than an Irish pence? Since it was fashionable for nineteenth-century Europeans to romanticize their real and imagined Celtic past, I wonder whether anyone told him that he shared blood with a race of warriors that conquered Rome.

The ancient Celts were a violent, tree-worshiping people who achieved high levels of skill as metallurgists, artisans, and warriors. Because they left no written records, our understanding of them comes from the mass of physical evidence they left behind and the biased accounts of Greek and Roman writers who considered them barbarians. The word "Celt" itself is a misnomer, a modern invention that has fallen out of favor with scholars, in part because the Celts themselves apparently didn't use the term, and in part because the "Celts" actually comprised many different tribes that shared similar physical traits, produced distinctive forms of art, architecture, burial practices, and metalworking, and fought with swords. (But because the conventional name is familiar and more convenient in this broad history, I'll continue to use it here.) As the Celts spread throughout what is now Europe and the Middle East, their languages began to diverge and eventually disappeared except for a few living tongues such as Irish, Welsh, and Breton, still spoken in the western extremes of the Celtic world.[5]

The material culture that was shared by the Celts during the late Iron Age from 450 to the first century B.C.E. is called La Tène, named after a Swiss village on the north shore of Lac du Neuchatel. In 1857, after the water level dropped during a drought, forty iron swords and other weapons were dredged up from the shallows near the shore. It's assumed that they had been plunged into the water in supplication to the Unseen Force, a common rite among the Celts, whose armaments have been found in the Thames (the most famous of these was the bronze-and-glass Battersea Shield), in a lake near Toulouse in southwestern France, and in the Brue River near Glastonbury, reputedly the burying place of King Arthur. (The legend of Excalibur, King Arthur's magic sword, is based on this practice.) While it seems that the only thing left to prove about Arthur is that he existed, weapons unearthed primarily from burial sites show that murderous swordplay was one of the reasons La Tène spread so far and so quickly from its beginnings in what is now Bavaria and Austria. This is despite the fact that the people in these graves were in life elite members of various selfish and disunited political units that sometimes fought among themselves, not players in a monolithic empire with a single goal. The centrality of the sword in Celtic life is evidenced by the image of the severed head, a revered object commonly inscribed on ceramics, helmets, coins, and shields. The head was a prized trophy in battle because it was believed to be the vessel of the soul, and that once it was hacked from an enemy's body it had powers of prophecy, speech, healing, the restoration of fertility, and independent movement.[6]

After a warrior harvested an enemy's head with a stroke of his sword, he liked to show it off. He might parade around with the head draped from the neck of his horse, nail it above his front door, impale it in the front yard on a pike, or mount it on a wall in the round, thatched houses the Celts built. Sometimes heads were embalmed in cedar oil, stored in chests, and displayed on special occasions for the enjoyment of guests. The heads of the most respected enemies were treated with extra consideration. After the top of the head was lopped off and the flesh removed by maggots and boiling, the skull was gilded and used as a chalice. The features of heads carved on Celtic coins are often distorted, with huge, vacant eyes, wild hair, horns, and tattoos. Some heads have two faces or even three. Smaller heads might be chained to a larger, central head. Typical Celtic stone heads found around the Hadrian's Wall region of northern England dating from the pre-Christian era are carved with wide, staring eyes and idiotic grins.[7]

When iron smelting and forging techniques spread into Europe from the Sinai Peninsula, there was an explosion of technology that made all this joyous mayhem possible, and eventually powered the advance of La Tène culture all the way to Turkey and Ireland. During the Bronze Age formidable weapons were forged from an alloy composed of 10 percent tin and 90 percent copper, which produced a bronze that was harder and more durable than either of these metals. To make the bronze even harder the Celtic smiths pounded on it. But because tin rarely occurs in the same ore deposits as copper it had to be imported. England had sizable reserves, but after those were depleted scholars conjecture that the commodity was brought in from the Near East. When that supply dried up, as well, possibly because of the disruption of trade routes, Celtic smiths turned to iron, one of the most commonly occurring metals in the earth's crust. This was not an abrupt transition—Celts had already been smelting iron in limited quantities. But although hammered bronze is harder than iron in its pure state, someone stumbled on the fact that when you inject a little carbon into iron when smelting it, then plunge that hot piece of metal into water, then heat it again, and finally pound on it, you can alter its crystal structure to make it harder than bronze.[8]

The fuel used in many Celtic and Roman smelters, and those across Europe well into the Middle Ages, was charcoal made from hardwoods such as oak, beech, apple, and hawthorn. Of these fuels hawthorn has the highest specific gravity, or density, and burns the hottest. Excavations of an eighth-century smelter in Spain revealed that most of the charcoal used to produce iron came from trees in the rose family, such as domesticated apple and pear and wild hawthorn. Researchers speculated that this wood was harvested by pruning, suggesting that it was taken from orchards and hawthorn hedges that had been encouraged to flourish by the removal of trees with a taller canopy in order to give the understory more sunlight. Pruning encouraged new growth and produced a sustainable supply of charcoal. In an iron smelter excavated in England dating to around the beginning of the Common Era, four samples of oak charcoal were found alongside a sample of charcoal made from hawthorn. Dating from about the same time period was an iron-making site in County Galway, Ireland, containing charcoal with a high percentage of hawthorn.[9]

When the Celts came to depend on their horses to such a degree that horse-worship became a cult—at about the same time their heartland was penetrated by Germanic tribes—the logical extension for the cavalry was the chariot. The first of these were built with four wheels; the later, two-wheeled models were much more maneuverable. A warrior standing in a chariot needed a longer sword so he could lean out to attack the head and shoulders of an enemy cavalryman or infantryman. Early La Tène swords were rather short, only twenty-two inches long or so. When the Celtic tribes in what is now France were finally defeated by the Romans their swords had stretched to three feet.[10]

The pinnacle of the battlefield achievements of La Tène culture was personified by a warrior chief named Brennus. (This might not have been his name but rather his title, thought to be derived from the Celtic word for "chieftain" or "strongman.") Brennus was a military leader of the Senones, one of some three dozen Celtic tribes that infested Gaul, a territory that included present-day France, Belgium, and Switzerland. Pressured by Germanic people pushing into their homeland, some of these tribes migrated into northern Italy, where they took up farming and cattle ranching in the Po River valley. But the Senones preferred to steal what they needed. Around four hundred B.C.E. they crossed the Alps and began raiding the Umbrians, who were eventually forced to abandon their hilltop towns and flee. The Senones established their capital at what is now Senigallia, 125 miles north of Rome on the Adriatic Coast. Then they turned their attention to their next conquest.[11]

They soon settled on Clusium, a site now occupied by the gorgeous hilltop town of Chiusi, in Tuscany, about forty miles southwest of Senigallia. According to the accounts of Titus Livy, a Roman historian, after Brennus began his siege the town appealed for help to Rome. Although the cities were neither allies nor friends, Rome responded by sending three envoys to negotiate a peace with Brennus. After the Celts accused Rome's representatives of being spies assigned to assess their troop strength, the meeting turned ugly. Insults were exchanged. There were unwholesome gestures. A skirmish erupted in which a chieftain for the Senones was killed by Quintus Fabius, a member of a powerful patrician family. When Rome refused to turn over Fabius to Senonesian justice,

Brennus abandoned his campaign against Clusium and headed off in a snit toward the capital.[12]

A disciplined and well-armed Roman army commanded by Quintus Sulpicius marched forth to defend the city. On 16 July 390 (or 18 July 396, according to other sources) they mustered eleven miles north of Rome around a little tributary of the Tiber called the Allia. What they confronted must have dismayed them. Not only were the Romans grossly outnumbered, twenty-four thousand "savages" against twelve thousand Romans, these unlettered "barbarians" were physically much larger than their adversaries. While the average Roman soldier, with his dark, rich skin, brown eyes, and glossy black hair, was five feet seven inches tall, the Celts, with their blue eyes, pale white skin, and cascades of fair and red hair, beards, and mustaches, stood six feet.

The Celts began whipping themselves into a frenzy, shouting insults, singing, thumping their swords on their shields in a rhythmic clamor, and whooping like demons. There was a bleating of horns, and a pounding of drums. Horses reared and whinnied. Some of the invaders acted out their version of the fate of the Romans, lunging at one another in mock battles, leaping, dancing, screeching, holding their heads and chests in agony, grasping their skulls by their long locks as they made slashing motions across their own throats, then collapsing. The highly trained and regimented Roman soldiers had never seen anything like it. As they glanced in confusion at their comrades right and left down the phalanx they must have wondered from what terrifying world this fever dream had emerged.

Finally, some of the Celts, having driven themselves insane, tore off whatever chain mail, armor, and helmets they were wearing and flung them aside. Chanting oaths, they raised their swords and their shields and stood before the enemy buck-naked, painted with pagan symbols, antic and horrible.

And then they charged.

The Senones drove point blank into the front lines of the Romans, with thundering chariots manned by two warriors (one to drive, the other to heave javelins and slash at the enemy with his long steel sword) and mounted warriors, two men to a horse (one to pilot the beast and the other to kill the enemy). Almost immediately the force of the attack shattered the Roman phalanx. When the Celts turned Rome's right flank, where the less experienced and more poorly equipped soldiers were posted, and drove them into the Tiber, Sulpicius ordered a hasty retreat. The battle lasted ten minutes.

What was left of Rome's six legions fled back to Rome, where a garrison locked itself into the heavily fortified Capitoline Hill. Although it would have made no difference to the outcome, they forgot to shut the gates of the city. Brennus' forces pursued them and poured into the streets of Rome, slaughtering civilians as they tried to flee. Then the looting and pillaging began. For seven months Brennus terrorized Rome and the valley of the Tiber, burning books and destroying almost all of Rome's records. Although his repeated assaults on the Capitol failed, the siege was causing the beleaguered Roman garrison to run out of food and supplies.

Finally, the Romans bribed the barbarians to leave. It has been speculated that the Celts were suffering from diseases caused by rotting bodies piled up in the streets. It is also possible that Senone settlements to the north had come under attack from other Italian tribes. Or maybe they just got bored by Rome and their siege and wanted to move on to fresh plunder. Whatever the reason, Brennus agreed to march away in exchange for a pile of gold.

The Roman historian Titus Livy described the negotiations: "Quintus Sulpicius conferred with the Gallic chieftain Brennus and together they agreed upon the price, one thousand pounds' weight of gold. Insult was added to what was already sufficiently disgraceful, for the weights which the Gauls brought for weighing the metal were heavier than standard, and when the Roman commander objected the insolent barbarian flung his sword into the scale, saying 'Woe to the vanquished!'"[13] It was a banner year for heads.

In his imaginative 1893 painting of the scene, the French artist Paul Jamin depicted Brennus as a hulking brute poised in a pool of blood on a doorstep, the elegant stone buildings of Rome behind him, the door held open by a cowering little man we presume is a Roman senator. On the floor of the room are five buxom women, some bound, all terrified, brought there for the conqueror's pleasure. They sprawl among chests of fine fabrics, coins, jewels, golden urns, and the severed heads of the losers. Brennus has tied his long red hair in pigtails. His mustache flows from his face down to his neck. He is wearing leggings and a leather tunic decorated with bronze or gold studs. He is grinning.

Jamin's lurid image is historically accurate only in respect to the sword strapped in a bronze scabbard at Brennus' waist, his domed iron helmet, and the bronze spear he brandishes. The painter based these representations on actual Celt implements of war he had seen in museums such as the Musée des antiquités nationales in Paris. By the late 1800s, thousands of swords, spears, javelins, shields, helmets, chariots, trumpets, a wide array of tools, and collections of horse tack had been unearthed in archaeological sites all over Europe and the Middle East and shared with the public. Along with hill forts, the ruins of settlements, and exhumed burial sites, they revealed the wide swatch of the world where Celtic tribes had penetrated and then settled. The spread of these bellicose people was finally halted in the first century B.C.E. by Rome, which had had been a second-rate military power when it was sacked by Brennus, but learned much from the Celts about the tools and art of war.

I decided to try my hand at making a sword like the one Brennus carried. I wouldn't simply buy a length of steel and sharpen it—What fun would that be? Instead, I'd make the steel from scratch, using the methods Celtic blacksmiths began employing three thousand years ago. One shiny afternoon in March, after the snow had melted and before the river flooded with runoff from the mountains, I drove to our rural fire station and applied for a burning permit. When the fireman asked me what I wanted to burn I told him I was planning to cook charcoal, and then use this charcoal to smelt iron. He stared at me. Then he wrote down "lawn clippings" and handed me the permit. "Make sure you have plenty of water handy."

Archaeologists have unearthed Celtic smelting sites all across Europe and the Middle East and reconstructed the primitive equipment and processes my distant ancestors used to turn iron ore into iron. Among the findings at some of these sites are hunks of charcoal made from hawthorn. Smelting iron was not a high-tech industry requiring the enormous complicated furnaces, rarified alloy metals, and massive consumption of energy employed in modern mills. Making iron in 2500 B.C.E. was a lot like baking bread.

My first step was finding the fuel. For chemical reasons this must be charcoal. Along with oak and beech, the most common charcoal found in La Tène and Roman smelters in Britain and Ireland was made from hawthorn. Hawthorn burns hotter and longer than either oak or beech, and unlike conifers and other softwoods, it contains relatively small amounts of the resin that causes wood to burn quickly and spit sparks. Oak and beech don't grow at Dark Acres, but hawthorn we've got in abundance. Although only one of our wild hawthorns is dead, and I would never cut down a live one simply to satisfy my curiosity (I'd also have to wait months for the wood to cure), I figured I could harvest enough dead limbs from living trees to produce the amount of charcoal needed to smelt enough metal for a single short sword.

After a few afternoons of pleasant work in the groves, I'd collected half a cord of gnarled limbs and dead trunks. I covered this mound, called a clamp, with sand and straw, leaving an open central chamber for access, which I would later cover, and set the clamp on fire. I estimated that it would take two or three days for the clamp to burn down. The object was to cook the hawthorn in an environment that contained as little oxygen as possible in order to drive the water and other unwanted compounds from the wood while transforming it into an almost pure form of hot-burning carbon. There would be enough flammable gasses in the wood to keep the fire alive, and I could monitor the progress of the clamp by the color of the smoke seeping from it. When it had changed from gray and white to a hazy shade of blue the charcoal would be ready.

As the watch began, I collected the materials I needed to build the stove. This oven would be a replica of a Celtic furnace called a bloomery: a hollow clay cylinder three feet high, three feet in diameter at the base, and half that distance at the mouth, with walls eight or nine inches thick made by layering coils of clay on top of one another. The interior would be coated with sand to make the walls more resistant to melting. When it was completed this homely little shaft would resemble Devil's Tower in Wyoming.

Except for deposits found in meteorites, iron does not exist in a pure state on earth: it is always bound up in compounds containing oxygen or sulfur.[14] Iron is so abundant that you can collect enough to make a sword by dragging a magnet through sand. But I didn't have enough years left in my life to squander one of them collecting magnetic grit. Most iron is produced from ore. I managed to get my hands on fifteen pounds of hematite, a dense, gray and rust-colored rock whose chemical formula is Fe_2O_3. I finished molding the bloomery. And then I built a bonfire from pine and roasted my ore over it in order to drive out any trace of water the hematite might contain.

Finally, the big day arrived. My charcoal clamp had cooked the hawthorn to perfection. I had broken up the larger pieces into briquette-sized chunks about the same size as my roasted ore. And I had finished curing the bloomery by building a pine fire inside to make sure it would be strong enough for the task ahead. My tools were lined up in a parade: an anvil, a sledgehammer, a heavy carpenter's hammer, a pair of steel tongs, a shovel, a metal bar, and an old-fashioned bellows, which I bought in a pawn shop. I wore jeans, a work shirt, a fireproof apron, leather gloves, hiking boots, safety glasses, and a baseball cap.

When the fire in the bloomery was shooting flames fed by air sucked in through the two small holes I'd drilled into the base, I began ladling equal amounts of charcoal and hematite into the chimney with my shovel. Then I fanned the flames with the bellows. I hadn't expected to have any success making iron on my first try. But today I would enjoy some beginner's luck that would allow me to justify calling myself a bloomer. After a half hour of feeding and bellowing, the first rivulet of molten slag ran out of the bloomery and congealed into a glowing lump the size of my fist. Before it could cool I scooped it up with the shovel, took it to the anvil, held it in place with the forceps, and began banging on it with the hammer. As sparks flew, the ash and crumbly molten rock fell away, revealing a porous, incandescent iron biscuit. I pounded away until I'd flattened it into a small round pancake.

There were two simple chemical reactions happening inside my bloomery. Carbon monoxide given off by flaming hawthorn charcoal was stripping the oxygen from the ferrous oxide, producing carbon dioxide, which was streaming upward out of the furnace. And iron crystals, bonding with a trace of carbon from the charcoal, were being drawn by gravity toward the floor of the bloomery, which was pooling with molten rock. As the hot, malleable iron collected, it melded into a porous mass whose cavities filled with ash and molten slag. My distant ancestors thought this spongy mass resembled a bloom. I don't know why; it didn't look like any flower I'd ever seen.[15]

The Celts increased the strength of the metal they extracted from their bloomeries by hammering on it, folding it over, reheating it, quenching it in water, and then heating it again and pounding on it some more to produce what is called wrought ("worked") iron. (The most famous example of wrought iron is the Eiffel Tower.) These swords, now made of what is called mild steel, were much harder than iron blades, kept their edge longer, and flexed without breaking. When the Celts could no longer make enough bronze because their sources of the tin ore called cassiterite had dried up, they had no choice but to turn from bronze to iron. They continued to live in a lethal world where their lives depended on their weapons.

Meanwhile, the time had come to reap my reward. Using the steel bar, I pushed the bloomery onto its side. And behold. Here was a smoking, angry lump of hellish-looking iron and slag the size of a honeydew melon. I grabbed it with the tongs, carried it to the anvil, and began pounding on it with the sledgehammer. This was not artful work, but it had to be done. Sparks flew as I gradually drove out the slag and ash and flattened the lump into a plate thin enough to fold over on itself. When the metal cooled somewhat and became resistant to the sledgehammer, I reheated it. Then I gave it a baptism in the nearest slough and once more reheated it. Next, more pounding and heating, heating and pounding, as I coerced the ugly hunk into a shape that was beginning to resemble a sword more than it resembled a melon. Finally, exhausted, I ran out of light.

On the patio the next morning I heated the iron, grabbed a hammer, and got busy again. I had extracted two pounds of iron from three pounds of hematite, but this would be more than enough metal for my purposes. When my wrought-iron plate was thirty inches long, three inches wide, and half an inch thick I turned my attention to the task of flattening it further and shaping a round, five-inch tang on one end about an inch in diameter over which I'd attach a grip. Every hammer blow made the metal stronger, which meant I had to hit it with increasing force.

Act 3 was less violent. I set about reducing the width of my burgeoning weapon with whetstones. As I ground away at the edges, I narrowed one end of the sword to a rounded point. Then I cut a five-inch length of wood from a relatively straight hawthorn branch two inches in diameter, bored a hole through the center of it, applied glue to the tang, and tapped the grip into place.

People who make blades for a living wouldn't give my sword high marks. But I thought it was beautiful. I swung it around to get the feel of it. Then I took it to a fence line along which I'd planted a row of Colorado blue spruce. These fast-growing conifers had flourished for a few years, but one winter they suddenly died. My theory, which my wife, Kitty, ridicules, is that our hawthorns began sending forth ruinous vibes directed toward every tree at Dark Acres that wasn't a Montana native, a sort of Celtic reaction to everything that wasn't a Celt. How else to explain why our pampered weeping willow did well for a couple of years, then suddenly shed all its leaves and withered?

With one sweeping stroke I cut the first spruce in half. Fifteen strokes later the bodies of every tree lay on the ground.

It was good to be the Brennus.

Four centuries after the Senones plundered Rome the hawthorn appeared again in the arsenal of Europe's "barbarians," as they battled Julius Caesar. But this time their weapon wasn't a sword forged by charcoal made of hawthorn, it was a hedgerow made from the living tree.

FOUR

The Hedge Layers

There is more violence in an English hedgerow than in the meanest streets of a great city.

—P. D. JAMES

It was like playing capture the flag in a maze, except that the other side wanted to kill you. There was nothing in the civilian life of Curtis Grubb Culin III that could have prepared him for this lush, thorny nightmare. He grew up an only child in Cranford, New Jersey, and graduated from Cranford High in 1935, a member of the tennis and chess teams. He camped under the sky with his Boy Scout troop and loved to fly fish for trout. After graduation he went to work putting up window displays for Schenley Industries, a distiller and liquor distributor headquartered twenty miles away in New York City, which also employed his father.[1] Schenley's best-known label was Black Velvet Deluxe.

Just before Christmas in 1938, as Europe braced for another world war following Germany's invasion of Austria and its annexation of the Sudetenland, he joined the Essex Troop, which had originally been a Newark horsemen's club that marched in parades. After it was mustered into the New Jersey National Guard, the Essex rode against Pancho Villa in 1916 and saw action in France during World War I. Following the war, the Essex was re-designated the 102nd Cavalry Regiment. Early in 1941 it was activated and shipped off for training at Fort Jackson, South Carolina. A home movie filmed in grainy color shows the unit working with horses. But it would soon trade in its mounts for armored vehicles.[2]

Culin was taught how to wage war with a tank. Small and compact,

Great Britain and northern France, with (*insert*) the extent of the bocage of lower Normandy, June 1944 (map by the author)

he possessed the ideal physique for duty inside its cramped and dangerous quarters. In a photograph of Culin in his combat uniform, which included a perforated tanker's helmet sporting earflaps and headphones, his glare promises, *If you mess with me, pal, you'll do it at your own expense.*[3] After tank training, he sailed with the 102nd from New York to Liverpool, arriving in England in October 1942, part of the vast mobilization preparing for the invasion of Normandy.

For 160,000 Allied troops, 6 June 1944 marked the shocking end of a long wait for the unknown. But for Culin, now Sergeant Culin, a tank commander with F Troop of the 102nd Cavalry Reconnaissance Squadron, the wait was not over. Forty-eight hours after the first Americans went ashore on D-Day, the 102nd was still anchored a thousand yards off Omaha Beach because there was not yet room on the beachhead to accommodate all the armored vehicles the Allies intended to throw against Germany in the first tidal wave of the invasion. It was a jittery and noisy wait. Allied bombers dropped thousands of tons of mayhem on fortified German bunkers ringing the bluffs overlooking the beach. In return, the Luftwaffe strafed and bombed American positions, and German artillery fired at U.S. planes and naval vessels shuttling men and materiel to shore.[4]

Finally, some of the 102nd's soldiers were ordered into landing ships and transported to Omaha Beach. But not Sergeant Culin. F Troop and its tanks would not go ashore until the following morning. What he saw along the way would have unnerved anyone—burning ships listing in the water, demolished vehicles, disfigured corpses and body parts buzzing with flies littering the bloody sand, all overlaid with the reek of smoke and death smudging the briny air.

Only three of the fifty-one heavy Sherman tanks sent to Omaha Beach survived D-Day, many of them sinking in rough water after their flotation devices failed. But Culin's mount was a relatively light tank called an M5A1 Stuart, which was brought ashore in a landing craft. It was named after Civil War legend Jeb Stuart, the Confederate general who advanced the art of reconnaissance and the use of cavalry to screen the infantry. The Stuart tank's mobility made it useful for scouting enemy positions and providing covering fire for foot soldiers, hence the motto of the 102nd: "Show 'em the way."

On 10 June, Sergeant Culin and the other three men in his crew climbed into their tank, and the squadron headed out on its first assignment of the invasion, to clean up pockets of resistance in the beach area north of the town of Isigny-sur-Mer. Their first taste of combat resulted in the capture of seventy-four enemy soldiers, among whom were several White Russians who had been conscripted to fight for the Wehrmacht, the German armed forces. The next morning the troopers moved again, this time southeast to Caumont-l'Éventé. On the edge of the village they were ambushed by German snipers and machine-gunners and lost their first man.

On 14 June, and again three days later, F Troop fought its first armored battles, along the road between Bayeux and Saint-Lô. Several American tanks were destroyed, and men Sergeant Culin had known for years were wounded and killed. Because most of the roads in western Normandy passed through Saint-Lô it was key to the American plan to break out of northwest France and liberate Paris, less than two hundred miles east, a strategic and profoundly symbolic goal that would signal the end of Germany's grip on Europe and hasten an Allied victory. The generals expected to capture Saint-Lô by mid-June. The United States, they reasoned, had a sizable advantage in troop strength, air power, and supplies. But Saint-Lô would not fall until 18 July, at a staggering cost. It is estimated that American forces suffered a hundred thousand casualties, more than twenty thousand of those men killed in action.[5]

As Sergeant Culin's first week of combat turned into his second he began to grasp what the brass had overlooked—the U.S. Army was fighting two enemies in Normandy. The Germans were one. And the trees were the other.

Twenty-four hundred years ago a Celtic tribe called the Gauls pushed into Normandy from its homeland at the headwaters of the Danube, built elaborate towns walled with timbers, cleared woodlands so they could farm cereal, and began raising cattle, sheep, and pigs. In order to keep the animals away from the grain and to prevent the incessant Atlantic winds from blowing off their topsoil, they planted hedges. Over the centuries hedges came to dominate and define the flat landscape of the region like no other feature. Some of these hedgerows date to the Middle Ages or farther back, but most were planted during a period of widespread land enclosure, similar to what was going on in Ireland and Britain, that lasted for a century, beginning around 1750. The hedges were planted primarily for firewood, the only source of fuel once the common right was withdrawn to take wood from the few woodlands that were left. For that reason the hedgerows were called "the Peasant's Forest."[6] These were not the intimate hedges of Ireland and Britain but formidable organic ramparts, a mosaic of interconnected barriers enclosing small plots, about five hundred to the square mile. Packed earth and rock berms up to four feet high and four feet thick formed the foundations, with drainage ditches on either side. Trees, bushes, and vines grew from the berms in a dense thorny mass. It was impossible to see through these hedges, and impossible to see over them. The big trees included beech, ash, oak, and chestnut. Of the smaller trees and bushes flourishing in the hedgerows the most common species was *C. monogyna,* the one-seed hawthorn.[7] The roots of this intertwined floral web acted like rebar, making the berms as adamant as concrete. Orchards grew in many of the fields, supplying fruit for Normandy's famous liquors, such as the apple brandy Calvados. The fields were open at one or more of the corners to allow farmers access, and by 1944 ancient wagon trails that had sunk with use over the centuries wound narrowly between adjacent fields. In some places the canopies of neighboring hedgerows had grown together, turning the trails into tunnels.

The Gauls were defeated by Julius Caesar in 51 B.C.E., during Rome's campaign to secure its borders and exploit native resources such as timber. Five centuries later the Franks overran the region. Later in the millennium it was the repeated target of Viking marauders. Finally, the Normans invaded, lending it their name. What remained constant during these floods of humanity were the hedgerows.

One of the other tribes Caesar defeated was also fond of hedges. The Nervii, who occupied present-day Flanders, might have used them for their farms; they definitely used them for defense. "The Nervii," Caesar wrote, "because they were weak in cavalry (for not even at this time do they attend to it, but accomplish by their infantry whatever they can) in order that they might the more easily obstruct the cavalry of their neighbors if they came upon them for the purpose of plundering, having cut young trees, and bent them, by means of their numerous branches extending on to the sides, and the quick-briars and thorns springing up between them, had made these hedges present a fortification like a wall, through which it was not only impossible to enter, but even to penetrate with the eye."[8]

In 1944 the hedgerows spread across more than fifteen hundred square miles of Normandy, covering most of the Cotentin Peninsula and stretching as far as sixty miles southeast. This peculiar geography is called the bocage, French for "grove," and it lay directly between Omaha Beach and Saint-Lô. No pleasant pastoral scenes summoned by the word *grove* appeared to Sergeant Culin as he and his unit made their way into Normandy. The Germans had occupied Normandy for four years, and knew how to fight in the bocage. Infantry, artillery, and tanks would hide behind three of the four hedgerows boxing in a field, soldiers peering out from spy holes cut in the foliage, backed up by snipers sprawled on platforms built in the trees. When the Americans advanced into the field the enemy would mow them down. The Wehrmacht did not contest every field, but the ones it chose to defend became fortresses. It was the outnumbered Germans who controlled when and where combat would flare, and they commanded all the elements of surprise. (Just after the invasion, General Omar Bradley, commanding the U.S. First Army, called the bocage the "damndest country I've seen.")[9]

To see what the bocage looked like in 1944, the film *Saving Private Ryan* is the wrong place to look. Although it is a harrowing and memorable war story, only the opening scene set in the cemetery was filmed in Normandy. The locations for the rest of the movie were in England and Ireland. The hawthorn hedges depicted in the scenes just after Tom Hanks and his men set out into what was supposed to be the bocage to find Private Ryan are lovely and bucolic, hardly the impenetrable barriers that confronted Sergeant Culin. (They are also in bloom, which means that the footage was shot in early May, more than a month before the beginning of America's assault on Saint-Lô.)[10]

The hedgerows were as deadly for tanks as they were for the infantry. The Germans had planted mines and roadblocks in many of the wagon trails, which were so narrow in places even the boxy little Stuarts got stuck, easy targets for the enemy, who blasted them with bazookas.

Compounding the difficulty was the rain. In Normandy the summer of 1944 was the wettest since 1900, a deluge that turned the trails into quagmires and watered the bushes and trees, making the hedgerows even denser and thornier than normal. Tanks trying to smash through the hedgerows got stuck in clots of mud and roots, or lost a track, or took so long to get through that the Germans were alerted and only had to wait for the tank to emerge onto the field before opening fire. If the tank tried to bull over the top of one of the shorter hedgerows its lightly armored underbelly would be exposed to German sniper fire. What was desperately needed was something that could quickly cut a passage through the hedge.

Sergeant Culin and a trio of officers were ordered to find a solution. What they came up with was a simple arrangement of five large sharpened steel blades welded onto the front of the tank. The thing resembled the teeth of a shark. General Bradley was invited to see a demonstration of this crude device, which was christened the "Rhinoceros." The portion of the hedge the Sherman tank plowed into exploded in a melee of branches, thorns, and mud, and the tank powered through the gap to the other side. Bradley was so impressed with the "Culin hedgerow cutter" that he ordered the immediate construction and installation of as many as possible. Almost six hundred "Rhino tanks" fitted with cutters would soon rumble into service. To find enough steel the welders turned to the thousands of anti-tank obstacles the Germans had littered across Omaha Beach to interfere with the Allied landing. General Dwight D. Eisenhower wrote later that the GIs were "gleeful" at the thought that German hardware was being turned against the Wehrmacht.[11]

Complaints about hedgerows form something of a leitmotif in the military history of France. Charles the Bald, a ninth-century king of West Francia—territory that was roughly the equivalent of France today—spent a lot of time and money constructing massive wooden palisades and fortified bridges in order to consolidate his royal conquests, protect his realm from marauding Vikings, and remind his people of his sovereignty. In an edict issued in 864, he complained that without his permission some of his subjects were building private chateaux and fortifications and enclosing land with *haies et fertés*—living hedges fabricated from tightly woven hawthorns. The miscreants were ordered to pull down the unauthorized constructions. Officers who fought in the Pacific theater in World War II complained that combat in the bocage was as bloody as any in the Pacific island jungles. Why had the architects of D-Day, who had meticulously planned every other detail of the largest invasion in history, underestimated the difficulty of breaking through the hedgerows? Planners had access to aerial photographs of the bocage, but they may have assumed what they were seeing was similar to the relatively tame whitethorn hedges of Britain. If they discussed the terrain of lower Normandy with their French allies, they seem to have ignored the information. Maybe they assumed that with their military superiority they would soon rout the Germans and roar through Normandy. One officer summarized the surprise of U.S. senior staff after the invasion: "Although there had been some talk in the U.S. before D-Day about the hedgerows, none of us had really appreciated how difficult they would turn out to be."[12]

Sergeant Culin and the three officers who devised the hedge cutter were awarded the Legion of Merit for their contribution. But it was the window dresser from New Jersey who was immortalized in a 1961 speech by General Eisenhower: "There was a little sergeant. His name was Culin, and he had an idea. And his idea was that we could fasten knives, great big steel knives in front of these tanks, and as they came along they would cut off these banks right at ground level—they would go through on the level keel—would carry with themselves a little bit of camouflage for a while."[13] Although Culin himself did not take credit for the idea, which he attributed to a "Tennessee hillbilly named Roberts," heroes, whether real or confected, always emerge from the fog of war. The United States needed champions, and the generals were happy to supply them. The "little sergeant" had the same egalitarian appeal as "the little guy," the backbone of America.

While the hedge cutter has become a military legend, its significance is a topic of debate.[14] Some historians credit the innovation with ending the stalemate and saving thousands of lives by restoring battlefield mobility to the Americans. At the very least the Rhino tanks increased morale because the men believed in them. Other historians argue that the Rhino tanks were not used as extensively as legend suggests because they were not particularly effective. The United States bombarded the area before the attack on Saint-Lô, tearing up some of the fields. This made it difficult for the Rhino tanks to reach speeds fast enough to plow into the hedgerows with sufficient force to break through. And the hedge cutter was not the only weapon used against the hedgerows. The "salad fork"—a pair of wood prongs attached to a tank that drove holes into the berm, which were then packed with explosives and detonated—played an important part as well. The army also brought in a few tanks fitted with hydraulic bulldozer blades that cleared the way once a hole was blown in the hedge. But exploding ordnance in the berms could not do the whole job: there was not enough explosive.

Skeptics argue that other developments contributed more to the U.S. victory in Normandy than these anti-hedgerow weapons. These included tactics for coordinating armor with the infantry and the engineers who blew holes in the hedges—which led to the catch phrase "One squad, one tank, one field"—and better radio communication between tanks and foot soldiers. But Sergeant Culin's story had greater theatrical appeal than the various more plodding measures. After the First Army broke out of Normandy in late July, Allied forces thundered across northern France, driving the Germans before them. On 25 August, Sergeant Culin was one of the first Americans to enter liberated Paris. Still fighting, this time in the Hürtgen Forest in western Germany (the longest battle the U.S. Army has ever fought), he stepped on an antipersonnel mine while on patrol one night. The blast shattered his left leg, which had to be amputated. As he fell he hit another mine, which badly injured his right leg. But he recovered, and returned home to work for Schenley, and married. The Little Sergeant died in 1963, when he was only forty-eight years old.

Hedges have been part of human commerce in Europe for at least five thousand years. When Neolithic people found it more profitable to raise cereal crops than wander around with their livestock looking for grass, they lined the boundaries of their fields with hedges in order to control their animals and stake out their land. In some coastal areas of Great Britain, such as Cornwall, these "hedges" were simply perimeters of stones. Europe is strewn with assart hedges, lines of trees and bushes that are the surviving remnants of forests cleared for farmland.[15]

Stone Age farmers might also have constructed dry hedges from branches woven together. In time these dead barriers might even come to life. To anchor the hedge, some of the green branches, cut from trees with a stone ax or hatchet, would be thrust into the earth, where they would take root and begin growing. New plants would then move in, taking advantage of the protection and retention of moisture the hedge furnished. (At Dark Acres we've built a couple of debris hedges by piling up branches and driftwood from the river in places where tree roots make it too much work to dig holes for a post-and-rail fence. These have been invaded by raspberry, rose, and dogwood—although so far no hawthorns have harbored there—persuading the neighbors' horses not to trespass.)

The existence of woven Neolithic hedges is suggested by the remains of the wattle-and-daub huts these early farmers built across Europe, the Middle East, and along the Mississippi River. Wattle is a slat or pole structure interlaced with thin branches that were often split to make them more pliable. Daub is mud or clay spread over the woven surface and allowed to harden (lathe-and-plaster building is a technique that descends directly from this early architectural practice). The branches were probably harvested from trees that had been coppiced, amputated via a cut near the ground and angled so that the rainwater would run off, thus preventing rot. Shoots called suckers soon spring from buds in the stump and grow into spindly branches. These shoots make excellent wattle. But the main purpose of coppicing nowadays in places such as Normandy is to produce a lot of fast-growing, sustainable timber and firewood without having to replant. Not all species of tree can be coppiced—coppicing does not work with pines or evergreens, for example, and mature hawthorns do not coppice well. But in mild and wet climates such as Ireland's, a youngish coppiced hawthorn can be harvested in ten years, starting the growth cycle over again, and extending the life of the tree by resetting its aging process. A hedge that is regularly pruned, and pleached every fifteen to fifty years will survive indefinitely.[16]

A British naturalist, Max Hooper, has made a study of England's hedgerows. Using documents such as farm records, he determined the age of 227 hedges, and found that they ranged from 75 to 1,100 years old. He then tabulated the various species of tree and shrub in each hedge. What he discovered was a striking correlation between species diversity and the age of the hedge. The larger the number of species, the older the hedge, for as a hedge ages it increasingly becomes a haven for plants fleeing the stomachs of cattle and the plow. According to his estimates, known today as Hooper's rule, if there are five different species in a thirty-yard stretch of hedge, the hedge is approximately five centuries old. There are several exceptions to Hooper's rule, but it has been widely used to determine the age of Britain's hedgerows, which turn out to be much older than previously thought. One of the ancient hedgerows Hooper studied lies in Monk's Wood, sixty miles north of central London. Called Judith's Hedge, it was named after William the Conqueror's niece, Countess Judith, who came to prominence after she had her husband beheaded for high treason against her uncle. She took sole possession of Monk's Wood and planted her hedge in 1075; it is now older than most of the buildings in Britain.[17]

Stone Age farmers no doubt discovered that a live hedge is stronger than a dead one. If it is manicured properly to encourage foliage it grows bigger, stronger, and denser over time, while a dead hedge becomes thinner and weaker. When the farmers returned to a tree they had cut down and discovered that it had coppiced, they would have seen that a couple of lines of these bristling plants might make a good barrier. And in time they probably took this information a step farther. By weaving the branches into a living wattle from thorny plants such as hawthorns and blackthorns they eventually created a green wall that was an affective barrier against wildlife, farm animals, and enemies. This arcane craft would later be called hedge laying by the English and *plessage* by the French.

In theory, laying a hedge is simple. First, plant a row of hawthorns. Saplings—called "whips" or "quicks"—that are two or three years old and eighteen to twenty-four inches high can be obtained from a commercial nursery . (It is, of course, possible to grow *Crataegus* from seeds, but it takes time and patience because the embryos of most species are notoriously resistant to being awakened from their dormant state. Germination typically will not occur before eighteen months, though the process can be accelerated somewhat by mechanically filing off a bit of the hard, woody shell protecting the germ, called an endocarp. Commercial growers use sulfuric acid to reduce the shell, making sure to wash it off when the job is finished.) The whips should be planted in the spring, set nine inches apart in well-worked soil. For a denser hedge, plant two rows of trees, or even three, making sure that they are a couple feet apart. Then leave them alone for a few years.

In six years, in County Waterford, where the average annual precipitation is forty inches, saplings of *C. monogyna* might have reached a height of eight feet. A hedgerow of a native Montana species such as *C. douglasii,* planted at Dark Acres along one of our property lines, where we get only fourteen inches of precipitation a year, however, might take perhaps a dozen years to reach that height. (While it would be fun to lay a hedge I grew myself, I don't have that kind of time.) Other species can be interspersed to create a hedgerow, and for reasons of biodiversity this is preferable. In Britain and Ireland hedge plants include crabapple, hazel, wild cherry, elder, and one of hawthorn's kissing cousins in the Rosaceae family, blackthorn (the source of some nasty spikes and the big blue berries called sloes used to flavor gin). In Normandy, oak, beech, and ash have been added to hedgerows, which are used more as wind breaks and sources of firewood than as pens for livestock.

Once the trees have reached a good height, they need to be transformed into a living wall, for a true hedgerow functions not simply to provide privacy or repel intruders or as topiary that resembles, say, the Seven Dwarves but as something that is as essential to a farm as any implement. To shape the hedgerow you'll need a hatchet, a heavy mallet, pruning shears, and a knife called a billhook (pronounced "billuck"), which has a long handle and a point that curves around the cutting edge. While used for centuries to gather branches for firewood, the billhook was also the English common laborer's weapon of choice, as closely associated with the peasantry as the switchblade once was with the American street thug. You'll also need heavy work gloves, long sleeves, protective glasses, and sensible shoes. And it's a good idea to keep a first-aid kit handy.

To begin, take the billhook and prune away about three feet of branches and foliage from the ground up on each tree. Then make a cut a few inches up the trunk of the first tree, cutting almost entirely through

Begin with a row of bushy young hawthorns in late autumn,
eight to twelve feet tall.

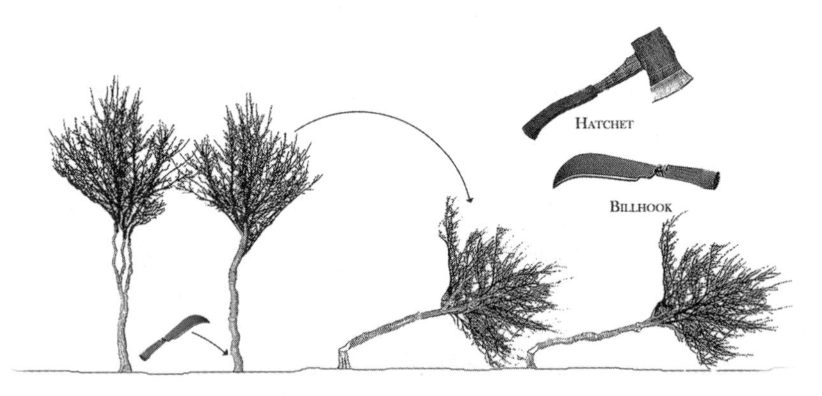

Trim the lower foliage, then make a deep cut near the base of the trunks with a billhook or a hatchet. The tree, now called a pleacher, should bend over at an angle of 30 to 60 degrees, with each pleacher resting on its neighbor.

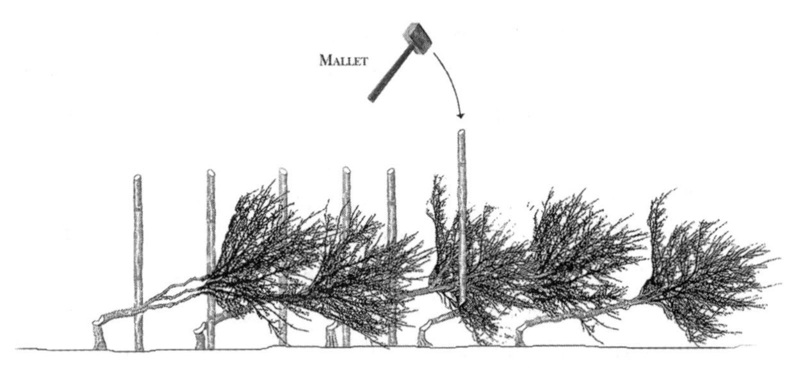

To give the hedge added vertical support, drive sharpened stakes into the row a cubit—about eighteen inches—apart.

THE ART OF HEDGE LAYING
(illustrations by the author)

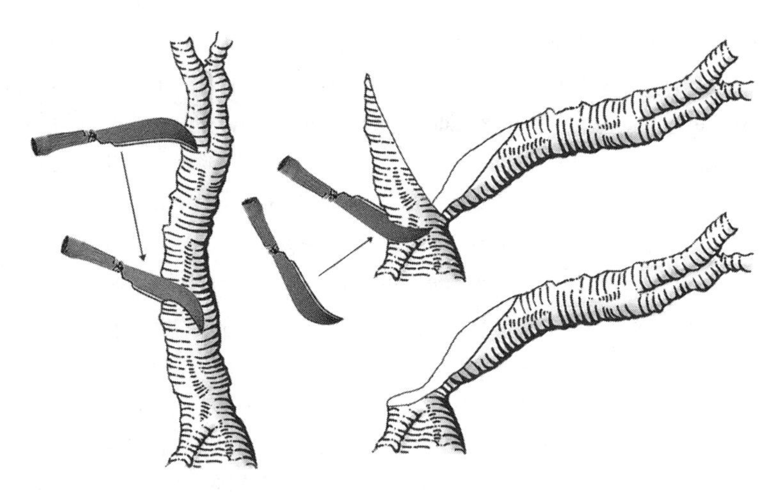

In step 2, use a billhook, hatchet, or small chainsaw to make an angled cut at the base of the tree, severing 90 percent of the trunk but sparing a flap of sapwood so nutrients can feed the rest of the tree. Remove the stub with another cut, leaving an angled surface that allows water to run off in order to discourage rot. Angle the pleachers upward to encourage the sap to rise.

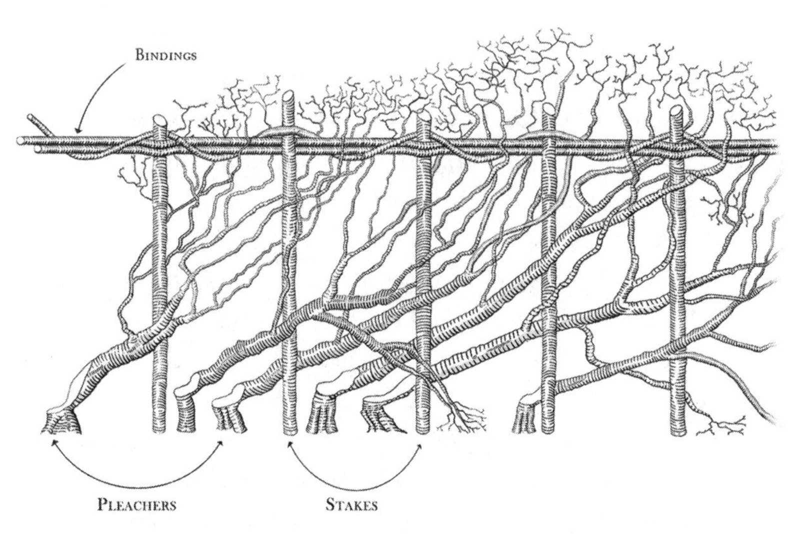

After the stakes have been pounded into the ground along the hedgerow, weave bindings made from thin greenwood "whips" around the tops of the stakes to give the hedge horizontal reinforcement. Then weave the branches of the pleachers around one another and the stakes.

the trunk and angling the slice acutely from high to low—the aim is to make a deep enough cut to allow the tree to bend at an angle but not deep enough to kill it. (With older hawthorns whose trunks are thicker you might need to use a hatchet or chainsaw for the first cut.) At the age of six an Irish *C. monogyna* will be three or four inches in diameter at its base, so there is little room for error, one reason why hedge laying is a skill that requires a considerable amount of practice. The process looks brutal, but the result is life-giving: hacking off the stub with the billhook or hatchet—the next step—promotes growth and prevents disease.

Now it is time to bend the first hawthorn and lay it sideways so that it rests on its upper branches at an angle of 30 to 60 degrees. (The bent tree is called a pleacher.) Because a portion of the bark and the sapwood have been left intact around the cut, the pleacher will not only continue growing, it will send forth a multitude of shoots that reach toward the sun. The more shoots, the thicker the hedge. And the dense foliage of a well-pleached hedge encourages inosculation, a phenomenon in which branches that touch sometimes wear away a bit of each other's bark as they sway in the wind and grow together in a natural graft, adding still further to the strength of the hedge. When the new growths reach a height of four or five feet, prune their upper reaches during the fall or winter. This will stimulate horizontal growth, compelling the pleachers to fill in gaps in the hedge. This result is a product of the anatomy and chemistry of the hawthorn.[18]

When a hawthorn seed germinates, it grows from only one spot, the terminal bud at the tip. As it gets longer the shoot develops nodes. Two side-by-side lateral buds will form from each node (*C. douglasii* will sometimes produce three buds). A thorn will grow from one of these buds and a shoot bearing leaves or flowers from the other. Like other woody plants, the terminal buds of a hawthorn contain a hormone called auxin that inhibits the production of shoots, thus enabling the tree to grow taller more quickly. Auxin not only controls the number of shoots that sprout but also the angle at which they jut out from main limb. If the terminal buds are removed, the tree will produce a plethora of new shoots. Pruning also spurs ramification—the divergence of the trunk into branches and smaller shoots—by removing the upper foliage so that the entire tree receives more sunlight. This is why hedges that are not regularly pruned fall into disrepair, developing gaps and growing taller instead of wider.[19]

Getting back to our hedgerow, once the trunks are pleached (another word for it is plashing, the art of training trees)—take some wooden stakes and sharpen them with your billhook. These stakes will give the hedge support while it is still in the growth stage, and must be stout enough to be driven into the ground under the pleachers with a mallet. Space the stakes a cubit apart. (The cubit—the distance between a man's elbow and the tip of his middle finger—was a unit of measurement in ancient Egypt and, according to Genesis, God's choice of rule for Noah's Ark. Hedge laying is the only discipline that still uses cubits.) The next stage is to weave the most pliable of the hawthorn branches through the stakes and the trunks to create a thick mat. Finally, twine several strands of long thin whips made of willow or coppiced hazel around the tops of the stakes, stretching them from one stake to the next all down the row. The method is the same as basket weaving, in which the whips, now called bindings, are the weavers. Eventually both will rot, but by then their job of anchoring the hedge will be finished. As the hedge ages and fills in, it will be necessary to prune it in order to encourage the dense growth at the base, which is what keeps farm animals from pushing through. The ideal shape for a hedgerow is a steep pyramid, wider at the bottom than at the top, so that all parts of the hedge receive sunlight.

In England regional factors have influenced the development of different kinds of laid hedges, resulting in some thirty styles of hedgerow. In beef country such as Oxfordshire, for example, the hedges need to be bullock-proof, so the stakes are braided with multiple binders for added strength. Hedges in the sheep and mixed-farming country around Derbyshire have no binders. Though hedge laying was once practiced all over Ireland, the craft died out, and all knowledge of regional styles along with it. But in recent years a revival in the art of the pleach has led to discussions among hedge layers of something called Irish Freestyle, which is a general blueprint using stakes with optional binders. Although new agricultural hedgerows are still planted in Europe, pleaching is directed mostly at mature hedges that have been neglected (left unpruned). If the branches of trees such as hawthorns are not trimmed regularly to remove dead wood and let sunlight foster the growth of the understory, they will become thin and develop gaps at the bottoms. A leggy hedgerow is of little use on the farm. The average pleach is performed on a hedge every fifteen to twenty years to make sure it remains a stock-proof (and people-proof) barrier. *Plessage* was once widespread in Normandy but these days it is rarely practiced, resulting in the decline in the vigor of bocage hedgerows. The hedgerows are now used primarily as a cheap source of fuel, and the trees are regularly coppiced so that they produce more wood.[20]

If you live in England or Ireland, and maintaining your hedge sounds like too much work, you can hire a professional. The Hedge Laying Association of Ireland and the National Hedge Laying Society in England offer lists of hedge layers whose skills have been certified by state-administered tests. Their fees vary depending on the condition of the hedge, but a rule of thumb is about £10 ($17) a yard. Depending on the condition of the hedge, a professional hedge layer can pleach as many as fifteen or twenty yards a day.

Neil Foulkes, who at the time of this writing was in his early fifties, began his career as a hedge layer in England in the 1980s when he took a class on wall building. "I was hoping to learn the craft of dry-stone walling, and had never even heard of hedgelaying." It was November, the season when trees are pruned, and he was told that class would start not with rock walls but with hedgerows. "From the first day I really enjoyed it." Once the class was over he continued his instruction by laying hedges under the tutelage of a professional. After he moved to Ireland in 1992 he planted and laid hedges on his smallholding. He was asked to lay a large hawthorn hedge around a "famine" graveyard, where people who died during the Potato Famine were buried. Foulkes has now worked in twenty-six of the island's thirty-two counties. One of his recent projects was restoring a hundred-yard hedge on the holiday farm of an English couple near Cootehall in County Roscommon. The hedge was an overgrown mix of *C. monogyna,* blackthorn (*Prunus spinosa*), hazel (*Corylus avellana*), willow (*Salix spp.*), holly (*Ilex aquifolium*), and spindle (*Euonymus europaeus*) that had not been pleached for two or three decades. It was growing on a steep slope that was eroding owing to Ireland's pervasive misty rain. It took Foulkes and his colleague eight days to finish the work, whereas a job in County Wicklow in which they laid the same length of a young blackthorn and field-maple hedge had taken only two days.[21]

"When I moved to Ireland," Foulkes told me, "very few people, including the farming community, had any idea what hedgelaying was." This ancient country craft was once a necessity routinely practiced by landowners all over Europe, but after World War II a widespread frenzy to achieve food self-sufficiency following years of war shortages and rationing led to the destruction of hedgerows to create larger fields. The amount of land in England planted to barley, for example, almost doubled in the 1950s. The larger tractors that were rushed to market demanded bigger fields unencumbered with hedges and trees. And the rash of new irrigation projects required drains that were free of roots. After 1962 farmers began bulldozing hedges— paid by the government under the Common Agricultural Policy (CAP) of what is now the European Union. This in a country of intimate rural landscapes that led Bill Bryson, an American author who served as head of the Campaign to Protect Rural England, to write later: "A least half the hedgerows in Britain predate the enclosure movement and perhaps as many as a fifth date back to Anglo-Saxon times. . . . The reason for saving them isn't because they have been there forever and ever, but because they clearly and unequivocally enhance the landscape. They are a central part of what makes England England. Without them, it would just be Indiana with steeples." In 1946, England was interlaced with some 300,000 to 500,000 miles of hedgerows. In the Republic of Ireland, where the government has authorized cutting down more native forest than in any other country in Europe, roads, housing developments and the restructuring of farm fields have led to the loss of thousands of miles of hedges in the past twenty years alone. Mismanagement and neglect have also taken their toll. It is estimated that in 2013 there were about 250,000 miles of Irish hedgerows whose predominant species was *C. monogyna,* which commonly grows on embankments that were created when drainage ditches were dug. Some of these hedgerows were constructed in the Bronze Age.[22]

In Normandy the hedgerows suffered enormous damage during the battle of Saint-Lô, but they were already falling into disrepair during the four years of German occupation, growing taller but losing density because no one was manicuring them.[23] Still, demographics and technology have changed the bocage more than bombs and hedgerow cutters. As people fled to the rebuilt cities for better-paying jobs, the traditional family farm was replaced by commercial, scientifically managed operations, the result of campaigning by the French National Institute for Agricultural Research. Bigger and better farm implements have allowed a mere 3 percent of France's population to feed the country, at times producing more food than France has needed. As the rural labor force shrank, fewer and fewer people were available to care for the hedges. As it does in the United Kingdom, CAP provides money for the modernization of French farms and offers farmers income support and crop subsidies. The cultivation of maize is now widespread, supplanting orchards and dairy cattle such as the Normande, whose fat-rich milk is used to make famous local cheeses such as Camembert. At the core of this brave new agriculture is the consolidation of land to create bigger farms, a geographic blunt force called the *remembrement.* Before the new policy, if a landowner bequeathed property to several heirs an individual might end up with a number of disparate hedged parcels. Now the mayor of a town has the power the rearrange an heir's holdings so that the parcels are contiguous. And to cash in on the trend toward bigger fields, farmers have started exchanging land with one another. The destruction of hedgerows that enables these neighboring fields to be joined continues to change the face of Normandy, where there are no regulations protecting agricultural hedges. By 2010, as a consequence, after the land consolidation that began in the 1960s, the inventory of hedgerows in Normandy had shrunk from 125,000 miles to 93,000.[24]

Preserving hedgerows is not simply a tree-hugger issue. After all, a hawthorn is not a tree most people want to hug. Without hedgerows much of Britain, Ireland, and Normandy would be as bare as the Great Plains because only a few patches of woodland such as Sherwood Forest remain intact. The removal of hedgerows has caused flooding along rivers, soil erosion, and the reduction of the surrogate habitat these linear forests provide for a wide range of flora and fauna. Both plants and animals use thorny corridors as safe havens. In England hedgerows have become the last refuge for six hundred species of plants, fifteen hundred species of insects, sixty-five species of birds, and twenty small mammal species. Conscientious farmers do not plow next to a hedgerow. They leave a thin margin of undisturbed soil, which encourages the growth of flowers, weeds, and grasses that feed a multitude of insects and beasts. Without this diversity of life, the crops and orchards that farmers have sown over the past half-century would never be pollinated.

A newly pleached hedge is strong enough to confine the most aggressive donkeys, horses, and cows. Of course, so is a barbed-wire fence, and it can be installed a lot more quickly and conveniently. Most Americans and many Europeans might ask, "So why bother with hedges?" But penning stock is not a laid hedge's only function. After even one summer's growth it will serve as a windbreak and a snow fence and provide protection against the elements for pasture animals, none of which can be accomplished by barbed wire. Hedges combat soil erosion and regulate the movement of water across the landscape. They reduce the amount of irrigation required in neighboring fields by blocking winds that evaporate water. They thus increase the value of a property by 5 to 20 percent. In Normandy hedges are still the only source of firewood, the cheapest fuel and a sustainable resource. And they are a critical element in flood control. Finally, if you agree with the climate scientists who blame humankind for global warming, you'll appreciate that unlike the production of barbed wire and the milled posts and poles and nails used to fabricate it into fences, on balance hawthorns *take* carbon from the atmosphere, they don't add it. For reasons of chemistry not all trees have this ability to scrub the air.

Ultraviolet rays in sunlight can trigger chemical reactions between atmospheric water and oxygen and volatile organic compounds (VOCs) that are formed from hydrocarbons and other substances released when coal and gasoline are burned. The result is the ozone found in smog. Scientists conducting a 2010 study sponsored by U.S. National Center for Atmospheric Research were surprised to discover that the globe's deciduous plants absorb 36 percent more VOCs than was previously believed. "Plants clean our air to a greater extent than we had realized," said the physicist Thomas Karl, the lead author of the study's findings. "They actively consume certain types of air pollution." In their study, Karl's team measured the VOCs absorbed by the black cottonwood, *Populus trichocarpa,* chosen because its genome was fully sequenced in 2006.

(Although some fruit trees in the rose family, such as cherry and peach, have been sequenced, no one has yet fully deciphered the genetic code for *Crataegus.*) In leaf experiments they found that when a black cottonwood is chemically or mechanically stressed—which in the natural world means compromised by pollution, insects, disease, or physical trauma—it produces higher levels of chemicals that deter threats in much the same way that white blood cells in humans are dispatched to combat infections. Because large amounts of these chemicals can also poison the tree, it produces more enzymes to convert them into harmless compounds, at the same time absorbing even more VOCs as it amps up its metabolism.[25]

Along with carbon dioxide, the chief greenhouse gas in the atmosphere, VOCs are inhaled through stomata on the surface of a tree's leaves, tiny pores through which oxygen is exhaled. Some trees themselves produce enormous quantities of VOCs, including isoprene and monoterpene, a quickly vaporized aromatic component of tree resin and the turpentine distilled from it. The bluish haze that gave the Great Smoky Mountains their English name comes from the fact that the air above them is filled on hot days with VOCs emitted from the spruce and fir forests below—a relatively harmless and radiation-reflecting airborne vapor, but a source of ozone if mixed with fossil fuel gases. When the planet was warmer trees evolved the ability to emit clouds of VOCs to keep themselves cool. Karl told me that these VOCs can be "detrimental to air quality in the presence of nitrogen oxides." To combat air pollution, "plant more trees, as long as they're the right trees," he advised. "This will reduce the levels of air toxics." The right trees include ash, birch, maple and Rosaceae species such as hawthorn, hackberry, pear, and peach. Don't plant maple, eucalyptus, oak, or black cottonwood because they emit more VOCs than they absorb. By way of comparison, in an hour on a hot day eucalyptus emits isoprene at a rate of seventy micrograms of carbon per gram of dry leaf, and monoterpenes at a rate of three micrograms per gram, while *Crataegus* emits none, while metabolizing VOCs from other plants and human-caused sources.[26]

Hedgerows can accomplish all these useful tasks only if they are maintained, regularly groomed, with the tops clipped to encourage the growth of the understory, and occasionally relaid to keep them from reverting to a row of trees. This reversion is what happened after World

War II, when most livestock growers abandoned old country practices such as hedge laying and replaced their thorn hedges, the barbed wire of Europe, with actual barbed wire.

Around the year 2000 European hedge countries began, haltingly, to recognize the value of what they were losing. A coalition of environmentalists, people who profit from tourism, and forward-thinking farmers started agitating for the preservation and restoration of the hedgerows. Today almost half the hedgerows in England are managed and protected, a conservation effort backed by law. The Hedgerows Regulations passed in 1997 prohibit farmers or governments from removing most countryside hedgerows without first notifying the local planning authority. These restrictions apply especially to hedgerows that are considered "important" because of their history, aesthetics, or benefits to wildlife. (Hooper's rule is used to determine the age of a hedgerow and thus one aspect of its "importance.") By 2010 some thirteen thousand miles of English hedgerows had been restored by vigorous pleaching. European Union governments now pay farmers if they sign a pledge to use fewer chemicals such as herbicides, and to maintain their hedges and the wildlife they harbor. Farmers are required to attend educational events, where they are presented with the case for hedgerows.

In the late winter of 2013 the cause in England was aided by Prince Charles, who was featured in a segment of *Countryfile,* a BBC television program devoted to rural life, as he laid a hedge on the royal estate at Highgrove in Gloucestershire. Wearing a ragged, patched gardening coat and wielding a billhook, the prince looked thoroughly professional. "When you first lay a hedge," he said, "if you do it well, it looks so marvellous and then the fun is to see three or four years later, it looks like a hedge that's always been there. I loved it. It's terrific exercise and at the same time it's a sort of hobby or interest to see if you can get better at doing it."[27]

At the urging of a conservation organization called Crann (Gaelic for "tree"), a dozen counties in Ireland have commissioned hedgerow surveys. Their goal is to determine the quantity, condition, and management of the hedgerows. Using representative samples, these inventories set a benchmark for determining the future health of this natural resource. Lawmakers acknowledged the environmental value of hedgerows, where two-thirds of Ireland's birds make their nests, when they passed the Irish Wildlife Act of 2000, which made it illegal to "cut, grub, burn or otherwise destroy any vegetation growing in a hedge during the nesting season of birds, from March 1 to August 31." Although this law is tremendously important to birds and the people who love them, it does not protect the hedges themselves for six months of the year. Many of Ireland's hawthorns were planted 250 years ago during the widespread enclosure of the countryside and some of them, nearing the end of their lives, ought to be replaced.

Since 1996 some fifty miles of hedgerows have been planted in Normandy by a conservation group called the Association de 50,000 Chênes (oaks), so named because William the Conqueror built his ships for the invasion of England in 1066 from oak trees felled in Normandy. But these new hedgerows are hardly keeping pace with those lost to bulldozers or neglect. Although there are thirty species of native shrubs and trees in the new hedgery, *C. monogyna* is rarely planted: European species of hawthorns and other members of the Rosaceae family are susceptible to a disease called fire blight. It originated in North America, and was first observed in France in 1972, but in 1985 it jumped from the hawthorns and attacked apple and pear orchards in Normandy.[28] Today it is illegal to plant a hawthorn that has not been grafted onto root stock that is resistant to the disease, and farmers must go through a laborious process to get a permit. As a consequence, the hawthorn, which has played a central role in the history of Normandy and been a benefit to humankind for millennia, now depends almost entirely on birds and animals for its continued existence.

As the English colonized the worlds that were new to them they brought along their love of gin, their contempt for the natives, and their diseases. They also introduced aboriginals to the notion that land could be owned by individuals. In order to announce this appropriation they grew hedges to mark the boundary between private property and common land, both of which were no longer available to the original inhabitants.

In 1804 the English established their first permanent settlement in Tasmania at a location that would become the capital city of Hobart.

The pioneers, who were largely prisoners and their guards along with a few free settlers, eventually ran out of food and were forced to hunt kangaroos. When the kangaroos ran out they began planting crops, surrounding the fields with "wicker" fences, or quick-set hedges created by planting *C. monogyna* seedlings shipped from the mother country, and then interlacing the branches into "wands" as the trees matured.

By 1830 hawthorn hedges lined wheat fields and sheep pastures all over the fertile Midlands region of the island. By 1840 wags were fond of saying that parts of the colony looked more English than England. An English writer and illustrator, Louisa Anne Meredith, who lived for eleven years in Tasmania, noted in 1852 that the most beautiful things she saw in her time there were the hawthorn hedges that reminded her of home and were so unlike the "hideous" deadwood fences she had seen in New South Wales.[29]

In the 1880s sheepmen and cattlemen abandoned their hedges because they found it easier to confine their beasts with a new innovation from America: barbed wire, which was less expensive in terms of the cost of labor and upkeep, and seemed more "modern." The vast hawthorn hedgerows of Tasmania went feral. Untrimmed and unmanaged, they morphed into simple rows of trees with gaps in their lower reaches that allowed animals to come and go as they pleased. Yet many of these historic hedgerows are still alive. The state has documented eighteen hundred miles of hedges that were planted along roads, and several times that number that make up internal farm hedges, albeit in various states of disrepair. Many large landholdings in Tasmania are still interlaced with six or more miles of hedgerow.

No one paid much attention to this forgotten forest of hawthorns until the 1960s, when a patrolman for the Department of Main Roads named Jack Cashion turned his pruning shears on orphaned trees lining the Midlands Highway near the town of Oatlands. His interest was not in restoring farm hedges but in creating topiary. The result was a roadside menagerie that included a crocodile, a reindeer, and a dinosaur. After Cashion died the topiary fell into disrepair. But a Parks and Gardens worker, Bradley Stevenson, began restoring it and adding new pieces. He explained the procedure to a reporter: "Being hawthorns, they've got large spines and they link together and hold themselves together.

. . . And once it's like that, because you've got a lot of these branches pulled over in a horizontal position they'll sprout out new branches and thicken up a bit." After the Midlands Highway was rebuilt in the 1980s to bypass the center of Oatlands, the town languished. As a civic project and to attract visitors, the community responded with an outpouring of topiary—including an emu, a giraffe, swans, and a golfer—designed by a Tasmanian sculptor named Stephen Walker and crafted by local people.[30]

While there has not been a significant revival of interest in living fences on the island—unlike the rebirth of hedge laying in Europe—a few landowners have begun to hire professionals to restore their hedgerows. Choosing a contractor for this work is easy because there are only a handful of hedge layers in Tasmania. One of them, James Boxhall, was flown in by a Tasmanian farmer to reinvigorate hedgerows on his property that had been planted in 1842. Boxhall became interested in the craft after finding some old tools that had belonged to a long-departed hedger. His fascination with the process and its creations led him to travel to England in 2007 and compete in a championship sponsored by the National Hedge Laying Society. His most recent work was laying a *C. monogyna* hedge a hundred yards long at West Park on the Dairy Plains of northern Tasmania. "It was coppiced six years ago due to it being gappy and full of old decaying wood," he told me. "Because of the coppicing it was really good to lay. Most stems were three inches in diameter and about 15 feet tall." He also shared his growing expertise with Kate Ellis on the mainland, who may be the only full-time female hedge layer in the southern hemisphere. Ellis, who holds a Ph.D. in ecology from the University of Saint Andrews in Scotland, has been commissioned to lay an overgrown, eighteen-hundred-yard hawthorn hedge at the Kyneton Botanic Gardens north of Melbourne.[31]

Opportunities for hedge layers in Australia are expanding, but working there requires dealing with a problem that does not exist in Ireland or England. Although *C. monogyna* is the most common hedging plant, along with blackthorn in some areas, it is considered an invasive weed that poses hazards for native flora because of its tendency to form dense thickets that dominate the understories of woodlands, restrict access, and harbor animals the Aussies consider vermin, such as rabbits.

In their efforts to achieve biosecurity the states of Victoria and South Australia have gone so far as to declare *C. monogyna* a Noxious Weed: not only are landowners restricted from planting it, if it is on their property they are required to destroy it at their own expense, grubbing out seedlings or killing them with herbicides. (Australians have good reason to be concerned about invasive species. One of the most destructive and bizarre of these is the cane toad, *Rhinella marina,* imported in 1935 from Hawaii because it likes to eat cane beetles, which were destroying sugar cane crops in the state of Queensland. Because there are no native predators immune to the poison contained in glands on the back of the beast's head, the toad population exploded. Millions of these four-pound toads now clog roads and yards, poison wild and domestic animals, and deplete the food supply for native insectivores. And there is no evidence that the cane toad has had any effect on the infestation of cane beetles.)[32]

In addition to Tasmania, English immigrants took the motherland's love affair with the hedge all over the globe. By the time New Zealand became a crown colony in 1841, *C. monogyna* had been extensively planted on the islands for farm hedges used to confine and shelter sheep. But hawthorn has no defenses against pests and diseases festering in the climate of coastal New Zealand, which is warmer than that of England or Ireland, and the hedges tended to become leggy, allowing sheep to wander. Although hawthorn was abandoned in favor of other trees, enough had been planted to make it a pest, which earned it a place in New Zealand's version of the Noxious Weed Act. Across the Canterbury Plains on the south island, for example, there are some 180,000 miles of hedgerows and shelter belts grown from Monterey pine and Monterey cypress imported from California, ramparts of evergreens subdividing the flat and formerly treeless landscape into a system of grids. While they provide a formidable barrier against winds howling in from the Pacific Ocean, these tall, uniform, monotone fortresses make the Canterbury Plains looks like the English countryside on steroids.[33]

The British practice of using hedges to control people reached its bizarre zenith in India during the 1800s. Beginning in 1804 the East India Company established a series of customs houses on rivers and trade routes to check the smuggling of salt into the parts of the subcontinent that it controlled. The company was making a fortune taxing this commodity, a critical necessity for Indians whose vegetarian diets provided very little natural salt. By 1869 the company's porous system of checkpoints had been expanded and reinforced until what they called the Customs Line stretched across twenty-five hundred miles of India and what is now Pakistan. It was patrolled by fourteen thousand enlisted men and officers.

Because the tax was an onerous financial burden on most Indians, smugglers continued to slip through the line with loads of salt carried on their heads. Company executives decided to build a stronger barrier. Originally this was a dry hedge constructed of stacked branches cut from the many species of thorn that thrive in India's sweltering climate. In 1858 the crown took over direct management of British India from the company, but continued to extract the unpopular salt tax (it was not lifted until 1946). And the government carried on work at a frantic pace on the hedge barriers. The cost of labor and upkeep, and the vast tonnage of organic matter that had to be collected, convinced planners that what they needed was a living hedge, which could be planted once, encouraged to grow, and maintained by occasional pruning. Perhaps the idea for a green barrier came from daydreams about pastoral childhoods in England. Or maybe the idea was suggested when the dry hedge began to take root and sprout shoots from cuttings plunged into the ground. But what probably drove the hedge laying was a series of Enclosure Acts, passed from 1845 to 1882, that mandated the planting of hedgerows across Britain as a way to keep peasants out of what was previously common land.

By 1876 this Great Hedge of India fortified more than fifteen hundred miles of the Customs Line. Some four hundred miles consisted of green hedge, another three hundred of a mix of green and dry, five hundred of purely dry hedge, and six of stone wall. According to the commissioner of Inland Customs in his annual report of 1869–70, in places this living fence was ten to fourteen feet high and six to twelve feet thick. It was planted from spiky species such as babool, *Acacia catechu,* Indian plum, *Ziziphus jujuba,* carounda, *Carissa curonda,* and three species of prickly pear, *Opuntia.*[34]

This monstrous instrument of oppression and human misery had been forgotten in both Britain and the Subcontinent when in 1995 a book conservator named Roy Moxham bought an old memoir in a London bookshop written by a British civil servant who had been stationed in India. In it was a footnote that referred to a "Custom's line" that stretched across India. Doubting that such a line actually existed, Moxham nevertheless launched himself on a quest to find out whether the story was true. After three years of travel and research he finally came across a clump of acacia and Indian plum growing on an embankment—the last remnant of the Great Hedge of India.[35]

As I read Moxham's book about what he called his "ridiculous obsession" I began to wonder whether the British and Irish had also brought their living fences to America. After poking around in nineteenth-century farm journals I was surprised to learn that, indeed, hedgerows were a common sight in the first decades of the republic. Because America is now a land of barbed wire, chain link, and brick walls, it may be hard to believe that hedges were once considered an agricultural necessity, as essential as a mule or a plow. They encouraged westward expansion by making it possible to farm the rich, deep earth of the prairies. But like most ideas brought from Europe to America, Americans had their own thoughts about the hedge and put their own spin on it.

FIVE

The American Thorn

So serious was the fencing problem west of Ohio and so completely did the possibility of growing the necessary fences on the soil captivate the public mind that the middle [eighteen] fifties witnessed what was justifiably described as a "hedge mania."

—CLARENCE H. DANHOF

George Washington's true passion was not revolution or government. It was farming. After he took sole possession of Mount Vernon in 1762 he devoted the eight-thousand-acre Virginia plantation to wheat, tobacco, and corn, and experimented with growing hemp, flax, cotton, and sixty other crops, an astounding effort when you consider that he was interrupted by a war and the administration of the new United States. He even planted mulberry trees to feed silkworms. He was influenced by a "scientific" approach to agriculture and farming methods advocated by Jethro Tull, the English agriculturist, who believed that tiny specks of dirt nourished a plant by entering its roots, so soil needed to be pulverized by deep and repeated plowing, with crops planted in rows. In the matter of agriculture Washington emulated his adversary during the Revolution, King George III. Nicknamed "Farmer" George, the king tended experimental plots at Windsor, grew crops, and raised a flock of Merino sheep at the Royal Botanical Gardens in London, inviting the derision of his critics.[1]

Washington's aim was to throw off what he considered the bondage of financial reliance on a single crop, such as a tobacco. Throughout his great love affair with Mount Vernon he kept detailed records of his adventures in horticulture. In a 1795 letter to his farm manager, William Pearce, Washington brought up a related obsession: "There is nothing which has relation to my farms, not even the Crops of grain, that I am so solicitous about as getting my fields enclosed with live fences. I cannot too often, nor too strongly inculcate this doctrine upon you." Because Mount Vernon was running out of timber for rail fences he became fixated on the idea of hedges to keep his farm animals away from his crops. Washington's weekly correspondence with Pearce was filled with instructions about the cultivation of hedgerows. He even gave his manager a manual on growing hedges. The two experimented with honey locust, willow, Lombardy poplar, cedar, and a number of thorn varieties, including *C. monogyna,* the whitethorn. Washington's plan was to start his hedges with fast-growing willows and poplars; then, as the slower-growing locusts, cedars, and other trees matured, the hedge would thicken. Or so went the thinking.[2]

Less than a year after he instructed Pearce to start looking into hedgerows, Washington's crop of honey locusts died, and he wrote to Pearce that "it would seem I think as if I never should get forward in my plan of hedging."[3] His experiment with hawthorns was also disappointing. In 1794 Washington had ordered five thousand *C. monogyna* seedlings from England. They arrived late in the spring, and few survived once in the ground. Those that did were puny and did not thrive. While *C. monogyna* flourishes across the wet and mild parts of Europe, American summers are too hot and dry, and the winters, at least historically, are too cold. This is not to say that *C. monogyna* does not grow in the United States. On the contrary, in California, Oregon, and Washington State, as in Australia, it is considered an aggressive, invasive pest that tends to dominate any piece of ground where it takes root, creating a monoculture that pushes out native flora, inhibits the movement of wild animals, and threatens local hawthorns such as *C. douglasii,* which cannot compete as well as whitethorn for the attention of birds because its haws are smaller and apparently not as appealing.[4]

Despite his failure to grow lasting holly and hawthorn hedges, Washington persevered with two species of locust. In order to wall off the vineyard he referred to as his "fruit garden," he combined closely planted individuals of these species with a series of ditches and banks, remnants of which are visible to this day. The fruit garden was originally intended for grapes, but they did not flourish, so he planted orchard trees and grasses instead, and created a nursery where he grew plants for the hedgery.

Being a devotee of English methods—so much so that in the mid-1790s he tried to find accomplished English farmers to whom he could lease four of his farms—he surely must have known about pleaching. But no evidence has been found that he ever tried his hand at it, possibly because he could never raise the thorn trees that respond best to the practice. Although the hedges he used to encircle his fruit garden kept out most of his livestock, without pleaching he could not have discouraged a determined hog. Washington allowed his swine to run wild so they would fatten themselves on forage, but they confounded him by pushing their way through his hedges to root around in his fields. (Because they could not be confined they could not be counted. In his inventories of livestock he wrote, "Of hogs many, but as these run pretty much at large in the Woodland . . . the number is uncertain.")[5] It would be intriguing to think that the widespread problems in modern-day Virginia caused by feral hogs, which kill wildlife, damage crops, and carry diseases, started with Washington. But most planters of the era let their swine roam freely, in part because pigs like to eat snakes and their eggs and will even dine on venomous serpents with no ill effect.

An 1834 letter to the editor in the *Farmer's Register* from a planter recounts a horseback ride taken across the farms at Mount Vernon thirty-five years after Washington died. "A more widespread and perfect agricultural ruin could not be imagined; yet the monuments of the great mind that once ruled, are seen throughout. The ruins of capacious barns, and long extended hedges, seem proudly to boast that their master looked to the future." Sometime after 1914, at the urging of Charles Sprague Sargent, an influential naturalist and a champion of trees, 250 whitethorns were planted at Mount Vernon as an homage to the arboreal history of the site.[6] In his diaries Washington mentioned that he had planted another species of hawthorn there, the "Small-berried thorn." This is the Washington hawthorn, *C. phaenopyrum,* a native American whose scientific name was bestowed on it in 1883 and whose common name derives not from the president but from the fact that it was introduced from the capital into Pennsylvania for use as a hedge plant. *C. phaenopyrum* is a favorite of gardeners because it does not bloom until late spring or early summer, when almost every other tree has finished flowering. Its thorns are two to three inches long and extremely sharp. No member of the species grows today at Mount Vernon, and it is not known why Sargent did not suggest that it be planted. If Washington had persevered with this tough native of the steamy Southeast he might have realized his heart's desire of a truly impenetrable hedge. Unlike the whitethorn, whose leaves tend to fall off and whose branches die in hot, dry conditions, the Washington hawthorn can take the heat.[7]

Another prominent Virginian was also experimenting with thorn hedges. Thomas Jefferson asked a Georgetown nurseryman, Thomas Main, if he knew of a hardy plant he could cultivate into living walls that would grow well at Monticello, his 5,000-acre plantation sprawled across an 850-foot swell of land in central Virginia. In 1805 Main convinced the president that the "American hedge thorn" would fit the bill. It grew abundantly in the countryside around the Federal City, and Main believed it would also flourish on the "little mountain." Joseph Dougherty, Jefferson's factor in Washington, wrote in 1805 to tell the president that he had bought four thousand whips from Main, which had been sent to Monticello on the second day of spring. These plants were Washington hawthorns. They were used in what Jefferson called, with the grandiose capital letters of the era, the South Thorn Hedge, encircling part of the South Orchard and all of the North Orchard, which included the Vegetable Garden, where he raised produce for food and carried out numerous horticultural experiments.[8]

In the beginning Jefferson was enthusiastic. "As a thorn for hedges nothing has ever been seen comparable to it[;] certainly no thorn in England which I have ever seen makes a hedge any more to be compared to this than a log hut to a wall of freestone." In 1806 and 1807 he bought sixteen thousand six-inch seedlings from Main and planted them six inches apart in the hedgery. But Jefferson's dream, like Washington's, was never realized. Documents from 1809 and 1811 indicate that most of the thorn hedge was abandoned. The problem this time was weeds. Jefferson wrote in 1807 to J. Philippe Reibelt, a native of France who hoped to establish a model farm in America: "The luxuriance of the soil by its constant reproduction of weeds of powerful growth & stature will bid defiance to the keeping your grounds in that clean state which the English gardens require."[9]

Jefferson gave up on hawthorn and planted a privet hedge around his Vegetable Garden and South Orchard. Although privet grows quickly and holds its own against weeds, Jefferson apparently abandoned it, as well. Maybe deer ate it, or jumped over it. Or maybe it was not dense enough to keep out smaller intruders such as rats. Whatever the reason, in 1808 he began planning a monumental barrier. This would become in time a ten-foot-high fence three-quarters of a mile long faced with a double row of pointed wooden pickets. It was too high for deer to jump over. And the pickets were placed "so near as not to let even a young hare in." The only trespassers this palisade could not shut out were schoolboys, who broke down the palings and swarmed into the orchard to heave fruit at one another.[10]

The hedging failures of the presidents did not dissuade their countrymen from planting their own. The future president of Harvard, Josiah Quincy III, removed seven miles of interior post-and-rail fencing from his sizable seashore farm at Quincy, Massachusetts, which he had inherited from his grandfather in 1784, and eventually circled the entire property with a hedge grown from *C. phaenopyrum.* This living fence, he said, saved him the time and money spent constantly repairing his deadwood barriers. Serving at the time as a Boston municipal judge but spending his summers at the farm, Quincy did not need to grow interior hedgerows to keep his cattle away from his crops because he never allowed the animals out of their stalls, an early experiment in the sort of inhumane industrial husbandry that has inspired so-called free-range farming.[11]

Because New England's soil is mostly rocky, fields had to be cleared before crops could be planted. Rather than hauling away tons of rocks, farmers did the obvious thing: they encircled their land with stone walls. But on Quincy's estate and along silty valleys such as the one along the Connecticut River, there was little stone, so farmers in the first decades of the nineteenth century built post-and-rail fences to protect their fields, and then began looking at hedges. While Quincy was publishing his gushing enthusiasm for his *C. phaenopyrum* hedge, some of the farm journals advocated instead the cockspur hawthorn, *C. crus-galli,* also called the Newcastle, and even *C. monogyna,* apparently unaware of Washington's failure with whitethorn or refusing to believe what they heard. *C. phaenopyrum* was criticized for growing too fast vertically and not throwing out strong lateral shoots. (Because of the multitude of English emigrants in America at the time, hedge laying must have been practiced on at least a few American farms, but written accounts of this Old World practice in the New World are rare.) In 1822 the *New England Farmer* complimented Quincy for his introduction of live fences. But the journal concluded that it would be many years before they would be extensively used. In 1813 Quincy received a letter from a Philadelphia judge, Richard Peters, who passed on a compliment from a friend who had visited the farm: "You deserve, not a crown of thorns, but a chaplet of hawthorn-blossoms, for daring to enclose your grounds with anything but the dreary stone fences which disgust a Pennsylvanian."[12]

Despite the accolades and the pleasure he took in working his estate, Quincy turned out to be an incompetent farmer. The only part of the enterprise that made any money was his salt-making operation. Although he was highly regarded as mayor of Boston and served as president of Harvard from 1829 to 1845, his political career on the national stage fizzled. In 1809 he proposed bringing impeachment proceedings against outgoing president Thomas Jefferson, whom he called "that snake in the grass." His motion was defeated in the U.S. House of Representatives, 117 to 1.[13]

In 1823, when his term as mayor began, Quincy leased the farm to a tenant who returned the cattle to their pasture. According to Quincy's biographer, his son Edmund, the emancipated beasts started eating the hedge, which had to be protected by rail fences to save it. "A hedge might be sufficient to restrain the wanderings of the civilized cattle of England," Edmund Quincy wrote, but these American cattle were bred in the woods and mountains of New Hampshire and Vermont, and regarded thorns as "a kind of *sauce piquant*—thrown in to increase the pleasure of the meal."[14] It was unlikely that Quincy's cattle savored inches-long, bacteria-infested thorns. (The camel is the only ruminant with a mouth tough enough to chew on the spikes of hawthorns, such as *C. sinaica,* the Sinai Thorn, which the beast might come across in Egypt.) But if Quincy's herd had exhausted the pasture, like all cattle they would have turned to the hedge, nibbling carefully on the leaves and the tips of the branches, pruning that would compel the tree to shoot out a frenzy of lateral branches, making it more robust over time, not less. Nonetheless, Quincy the elder was apparently convinced by his experiment that hawthorn hedges would not work in New England.

Still, voices in other parts of the young republic continued to urge American farmers to replace their fences with hedgerows. One of the most influential of these was that of a popular British pamphleteer named William Cobbett. Fearing that his controversial positions on the issues of the day would get him arrested for sedition again (he had been locked up from 1810 to 1812 in London's ghastly Newgate Prison), Cobbett fled to the United States in 1817 with his sons and began a two-year exile on a farm on Long Island. His experiences there led him to publish *The American Gardener,* one of the first works on horticulture in the United States, now considered a classic. In it Cobbett described the methods used to grow what the British call a "quickset" hedge—*quick* in this case meaning alive, as in "the quick and the dead."

During the first two weeks of October, he advised, take nursery seedlings of *C. monogyna* one, two, or preferably three years old, and prune their roots to a length of four inches. Plant a row of these seedlings in well-worked soil twelve inches apart. Then, six inches away from this row, plant another line of hawthorns, each seeding opposite the middle of the intervals of the seedlings in the first row. Come spring, cut the seedlings off a half-inch from the ground. This act of coppicing, Cobbett promised, would cause the hawthorn to send out shoots that would grow three or four feet in a single summer. Unlike Jefferson, Cobbett was aware that the row had to be kept weed free by constant hoeing. "Let the plants stand two summers and three winters," Cobbett advised, "and cut them all *close* down to the ground as you can in the spring, and the shoots will come out so thick and so strong, that you need never cut down any more."[15]

But they still must be pruned. Cobbett recommended that hedges be "ten feet high, and five feet through at the bottom, because then I have fence, shelter and shade." But even at five feet, the height it would reach in six years, the hedge would deter the boldest boy from stealing peaches or watermelons from the field it protected because of its dense thorns and branches. Another reason Cobbett believed that a hedge was superior to a fence was that it kept out poultry. Fowl seldom fly over a fence, he argued. They perch on it, then drop down on the other side to ravage the garden. But no chicken can perch on a thorny hedge.[16]

Cobbett calculated that an acre of orchard would require a hedge nine hundred feet long, which would cost fifty-three dollars over the course of six years, the equivalent of roughly a thousand dollars today. This is not much for what is arguably the most essential structure on a farm, and one that would theoretically last forever. America has "English gew-gaws, English Play-Actors, English Cards and English Dice and Billiards; English fooleries and English vices," he observed. Why not English hedges, as well, instead of those cheerless, sterile-looking enclosures Americans throw around their gardens and meadows?[17]

Three omissions in Cobbett's essay on hedging are surprising. First, he neglected until near the end to tell his readers that they would have to build a wooden fence around the hedge to protect it for at least six years from beasts such as rooting pigs and therefore did not include the cost of such a structure in his estimates. And since he had never actually planted a hedge, he could not state with certainty whether one would grow in America. But these lapses were in keeping with Cobbett's colorful character. This was the man who popularized the term "red herring" in an 1807 essay, in which he offered an apocryphal story about how as a boy he once distracted hounds chasing a hare by dragging this fish across their path (a herring turns red when it is cured in brine or smoked). And Cobbett is notorious for having dug up the bones of Thomas Paine in 1819 and shipping them to England in a trunk, claiming that his fellow Englishman and pamphleteer ought to be buried in grand style in his native ground. When Cobbett died twenty years later the trunk and its contents were still in his attic.[18]

Farmers in the mid-Atlantic states, however, apparently heeded Cobbett's call because many miles of hedgerows would be planted in the years to come. For the most part these were grown not from *C. monogyna* but from American thorns. One of the noted advocates of native hedging was Andrew Jackson Downing, a journalist and horticulturist who popularized the idea of a central park for New York City. Regarded as the father of American landscape artists, Downing acknowledged that

C. monogyna could be suitable for living fences in the United States—he himself owned a thousand feet of excellent hedgerow made of this species, and he reported that he had seen a mile of such "promising young hedge" growing on a Geneva, New York, farm. But he maintained that on the whole *C. monogyna* would not thrive in America. Americans, he argued, needed an American hedge. For many years he insisted that the best species for this were *C. crus-galli* and *C. phaenopyrum.*

"Fifteen years ago," Downing wrote in 1847, "a person riding through the lower part of New Jersey and Delaware, would have been struck with the numerous and beautiful hedges of Newcastle and Washington thorns. Whole districts, in some parts, were fenced with them, and nursery-men could scarcely supply the demand for young plants." The Newcastle produces reddish berries, copious thorns, and small, malodorous white flowers like most *Crataegus* species. It was named for the northernmost county of Delaware, where so many miles of the species were used as farm fencing in the first half of the nineteenth century that Newcastle was reputedly the most hedged county in America.[19]

But when these thorns began to die Downing was forced to change his mind. In less than three years "a very beautiful hedge of the Newcastle thorn" growing on the Hudson River had been almost completely ravaged. The culprit was the larvae of several species of apple-boring beetles that also feed on the sapwood of other trees in the Rosaceae family, such as pear, quince, and serviceberry. But Downing was caught up in what the press of the time called "hedge mania," a near-hysteria fueled by the discovery of the incredibly fertile soils of the prairies coupled with the realization that because of the lack of fieldstone or timber for fencing there was no economically feasible way to grow crops on these plains. Sticking with the central lessons learned by European hedge layers working with hawthorn, Downing became an advocate for two other trees that are also armed with spikes but are more resistant to insects and sun. First was the common buckthorn, *Rhamnus cathartica,* which was being imported from Europe to America as early as the turn of the nineteenth century. Its virtues as a hedge plant are its vicious thorns and the fact that it doesn't need to be pleached because its branches grow naturally interwoven.[20] Downing, who died in 1852 in a fire on a Hudson River steamboat at the age of thirty-six, did not live long enough to see the buckthorn's major flaw. It would become an outlaw flourishing in impenetrable thickets that forced out native species populating the understories of woodlands. At least six states have declared it a noxious weed, a legal classification that permits governments to ban and burn it.

Downing's second choice for American farm hedges was the Osage orange, *Maclura pomifera.* This is a strange tree, indeed. A native of the Red River valley in Texas and Oklahoma, in the chigger-infested cotton country where my father grew up during the Depression, it boasts wood that is immune to termites, rarely rots, and is so hard it resists being glued or machined. Some Osage oranges are male and some are female, which produce wrinkled fruit three to six inches in diameter weighing two or more pounds that resemble tiny orange brains. These may or may not be eaten by small mammals, according to which naturalist you believe, but humans find them inedible. Nevertheless, some people bring them home and put them under their beds because they believe that they contain an anti-fungal agent that chases away bugs. The Osage orange's qualifications as a farm hedge plant are its thorns, of course, as well as its ability to survive droughts that savage other trees and its enthusiastic response to pleaching.[21] (Downing did not live to see the realization of his other dream, New York's Central Park. Five species of hawthorn grow in the park, including *C. monogyna, C. crus-galli,* and *C. phaenopyrum,* but ironically there is no Osage orange and buckthorn was apparently never welcome.) Downing's enthusiasm for *Maclura pomifera* was influenced by a man who, as much as anyone else, made it possible for Americans to venture out of the East to try their hand at farming the treeless prairies.

Jonathan Baldwin Turner grew up in Massachusetts and graduated from Yale University in 1833 with a degree in classical literature. He immediately took a teaching job at a small denominational college in the wilds of Illinois. Two months later a cholera epidemic broke out, decimating villages up and down the Mississippi valley. Later that summer, seeking to put aside for awhile the stress and anxiety caused by this calamity, Turner and two other professors decided to explore the country by taking a horseback trip to Chicago, two hundred miles north. They could not find a guide but were allowed to travel with a party of defeated

Potawatomie Indians who were heading to Chicago after being forced to sell their homeland on Lake Michigan to the whites (land bought for three cents an acre from the vanquished and immediately sold for a hundred dollars an acre to the victors). There were almost no roads or bridges. In some places the prairie grass grew higher than a mounted man's head.[22]

What struck Turner was how far apart settlers on the plains lived from one another, and how threadbare were their farms. Schools and their "civilizing" influence could never be established in a place whose population was so scattered, he decided. And without schools how could this big, empty region become as cultivated as New England? What was needed were more people. But how to attract them? Turner's solution was a sublime leap of imagination. Good fences, he reasoned, would eventually make good education. It was not land or fertile soil that was a scarcity on the prairie, it was a way to keep livestock away from crops. A few farmers who had settled near one of the rare patches of woodland had found the time and material to build "worm" fences, the Virginia split-rail style of enclosure in which rails are laid on top of one another in a zig-zag pattern. Some others had enough resources to buy timber that was hauled in from far-off forests. The lifespan of these untreated worm fences was only eight to ten years, and they were prone to destruction by grass fires.[23] It was estimated that fencing on the prairie consumed a month of a farmer's labor each year, and that the product of this hard labor was worth more than the land. Since the majority of Illinois pioneers did not have the means to enclose gardens and fields they became stockmen, turning their animals out to fend for themselves. The result was a widespread chaos of unregulated grazing on public and private lands by herds of cattle and America's growing population of feral hogs. Any unenclosed land was treated as a commons, regardless of who owned it. An 1830 manual for farmers summed up the problem: "The mode of inclosing, as here practiced and the urged necessity there is for the strongest fortification, in consequence of the barbarous practice of suffering stock of all kinds to run wild, keeps the farmer poor, and groveling and ignorant, and creates more rustic quarrels than any other thing, whiskey not excepted."[24]

Turner pondered the conundrum: few farmers would move to a

place where crops were in constant jeopardy, and the prairie states would never have enough timber to protect them. Finally, it struck him that the solution to the Illinois farming problem was rooted in the history of his English ancestors: stock-proof hedges. Almost any species will grow on the prairie, where the rich soil is hundreds of feet deep in some places, and Turner reckoned that hedge plants ought to flourish there. The task would be to create a living fence that, as the wags said, was "horse high, bull strong, and hog tight." Turner may not have known about the dismal experiences Washington and Jefferson had with hedging, or he may have believed that he could do better. Perhaps he had seen one or more of the few successful English-style hedges planted on the eastern seaboard. So at considerable expense, especially on a teacher's salary, he ordered a load of *C. monogyna* from England. But he soon discovered that during prairie summers the whitethorn defoliated, just as it did in Virginia. He experimented with other spiky plants such as barberry, box and locust, but rejected them all. He also tried his hand at the American thorns, which were all the rage in the agricultural journals, but found them lacking, as well, probably because they did not grow quickly enough.[25]

Planting trees and studying them throughout the 1830s and 1840s, Turner finally seized on the Osage orange as the answer to the fencing problems of the heartland. In 1847 he released his first flyer advertising seedlings for sale. "One hedge around a farm secures orchards, fruit-yards, stables, sheepfolds, and pasture-grounds from all thieves, rogues, dogs, wolves, etc.," he gushed. For the first couple of years, prairie people called the orange Turner's Folly. But when it proved itself a true American farm hedge, tough, fast-growing and cheap, farmers rushed to plant it, especially after they saw that crops were healthier when they were protected from the dehydrating force of the wind. The railroads, which were required by law to keep free-roaming cattle off their tracks, planted hedges as well. *Maclura pomifera*'s use as a living fence spread over much of the United States. By 1895, seventy-two thousand miles of stock hedges had been planted in Kansas alone. Even easterners began planting Osage orange after lumber became too expensive. In addition to other plants that took refuge along fence lines, Osage orange spawned thousands of miles of volunteer hedges that redefined the countryside from New England to Texas.

After 1874, hedge mania on the prairie subsided, owing to the mass production of barbed wire, whose design was obviously inspired by the effective animal-controlling spikes of hawthorn and Osage orange. A few farmers continued to prefer living fences to metal, which rusts, and they planted and cultivated hedges made of both species. On the prairies, where fence posts had to be buried in the ground before wire could be strung, farmers stuck with what they knew, and used Osage orange because it resisted rot in an era when chemical treatments for wood were unknown. Posts that have been in the ground for half a century without showing signs of deterioration can still be found. As barbed wire fences began to crisscross the country, farmers and stockmen let their hedges fall into neglect. Without periodic trimming to promote the growth of foliage at the base, prairie hedgerows degenerated into crude borders of trees with massive crowns that grew together, creating tunnels around roads that had been hedged on either side. Landowners began bulldozing Osage orange groves to prevent them from spreading. But many miles of this picturesque feature of the prairies are still thriving. The most famous of the historic hedges was planted by a farmer who staked a claim to 160 acres of Nebraska tall grass soon after the 1862 Homestead Act was passed. Forming the southern boundary of what is now the Homestead National Monument is a half-mile of Osage orange hedgerow, which may have been pleached by its original owner. In the East, individual trees still survive from the era of hedge mania. An extension agent in Montgomery County, Pennsylvania, told me that people are always bringing her the fruit of the *Maclura,* curious to find out what it is.

Inspired by the story of Roy Oxham's search for the Great Hedge of India, I flew from Montana to Philadelphia one August to see whether I could find any remnants of farm hedges grown from the two species of hawthorn Downing once championed, or from Turner's favorite tree. As I looked down from the plane at the farm country of Pennsylvania I was struck by how much more the verdant terrain resembled the parts of Ireland I'd seen than it did of any of the two thousand miles of America I'd just been flying over. Not only do both Ireland and Pennsylvania get around forty inches of rainfall a year, which keeps them green; the landscape of rural Pennsylvania, like that of rural Ireland, appeared to be defined by hedgerows. Looking more closely, however, I saw that these seemed to be almost entirely assart hedges, the surviving fringes of wild maple, birch, and cherry forests that had been cleared for crops and pasture, not the planted, premeditated hedgerows of Ireland. There were certainly some individual hawthorns or groves visible—Pennsylvania's three native *Crataegus, C. crus-galli, C. phaenopyrum,* and *C. mollis,* the downy hawthorn—are common here.

I didn't expect to find any surviving hedgerows, however. Although hawthorns live a long time, are protected by spikes, and have extremely hard wood they are not immune to a farmer intent on grubbing them out with a saw, fire, poison, or a team of mules. But I thought that I might be able to find some mature *Crataegus* that had originally been planted for use as a living barrier. Because the area had been one of the most productive farmscapes in the world during the height of hedge mania in the 1850s, I decided to begin by poking around the western suburbs of Philadelphia, in particular the Main Line (named for the primary track of the Pennsylvania Railroad, completed in 1831). Obviously there wouldn't be any farm hedges lining the vast lawns of the Merion Cricket Club or gracing the campuses of Villanova or Bryn Mawr. But because the Main Line is where wealthy Philadelphia families once had their country homes, I guessed that one of these old estates might harbor the ghosts of hedges grown from *C. phaenopyrum* or *C. crus-galli,* species recommend by a writer from this very corner of the state in an 1852 issue of the *Pennsylvania Farm Journal.* After all, not that long ago—in the late 1880s—there were still more than fifteen hundred cattle grazing in the township of Merion alone, which is now home to many of the Main Line's leafy, affluent neighborhoods.[26]

The Ardrossan Estate in Radnor Township was cobbled together from a number of farms during the late 1800s by Robert Leaming Montgomery, an investment banker who loved fox hunting and wanted to live in the country so he could indulge his fascination with agriculture. The estate and its denizens were the inspiration for *The Philadelphia Story,* the witty 1939 Philip Barry play starring Katharine Hepburn. Once boasting a thousand acres of grain, alfalfa, corn, and other crops, Ardrossan has been subdivided and its herd of pure-bred Ayrshire dairy cattle sold. But still intact are 350 acres of lazy green hills, old stone farmhouses, and a fifty-room Georgian mansion. The estate's property manager, Ed Proudman, told me he was not aware of any hedges of the sort I described, but he gave me permission to go have a look for myself.

So on a luminous, overcast afternoon I began wandering around the estate, an intimate and bucolic landscape without a car or a power line in sight. When I passed a herd of Black Angus and a pasture lined with big round bales of hay it was easy to imagine the place as the working farmland of two centuries ago, rather than Montgomery's feudal shire, where the "peasants" lived on the same land as their lord. I came across several hedge-like structures, but these were merely dense lines of foliage that had flourished in the safety of fencing made from webbed metal and post and rail. The only hawthorn I found was a puny *C. crus-galli* recently planted under the window of an outbuilding.

I headed northwest, and in a couple of minutes entered Chanticleer, a stunning forty-seven-acre formal garden that was opened to the public in 1993. Like Ardrossan, this land had been farmed by Welsh-speaking Quakers beginning in the late 1600s. In 1912 it was bought by Adolph Rosengarten, who founded a company that would merge in 1927 with Merck, then the largest pharmaceutical corporation in the world. No, there aren't any old hawthorn hedges here, the receptionist told me, but there *are* three hawthorns at Chanticleer. These turned out to be mature *C. viridis,* the green hawthorn, growing on a terrace inside the manicured grounds of the house where the Rosengartens once lived. One of the horticulturists told me these specimens had been planted to replace the original *C. crus-galli* that once grew in the same spot.

Twenty miles south of Ardrossan, in New Castle County, Delaware, lies Winterthur, the thousand-acre property that had been the country home of Eleuthère Irénée du Pont de Nemours, founder of the gigantic chemical company. (DuPont started out in 1802 as a gunpowder manufactory, and the mills were still on the grounds.) Once sprawling across twenty-five hundred acres, in the early 1800s the farms raised grain, Merino sheep, cattle, and fruit trees. Du Pont's great-grandson Henry F. du Pont was a horticulturist who designed a world-class, two-hundred-acre formal garden at the estate. As at Chanticleer, every tree in this garden is catalogued and tracked. I found out from the "Winterthur Bloom Report" that the previous May a single *C. crus-galli* had blossomed in the garden, a lonely, pampered thorn that had certainly not belonged to a proletarian farm hedge.[27] In fact, the director of the garden, Chris Strand, told me there were no hawthorn hedges at Winterthur. Could there be some in other parts of New Castle County, which in the early 1800s was as hedged as any farmland in England? (Nearby Longwood Gardens, a thousand acres once owned by another du Pont, Pierre Samuel, has been an arboretum open to the public since 1798. Growing on the grounds are thirteen species and hybrids of *Crataegus,* but no farm hedges.) After a spokesman for the Delaware Center for Horticulture told me that his organization had no evidence that such hedges existed I gave up and decided to look elsewhere. But I didn't come across anything of interest in the way of hedges. I flew home, as empty-handed as Roy Moxham after all but his ultimate trip to India to search for the Customs Hedge.

If I had been as obsessed as Moxham I would have persevered and discovered two ghosts at Winterthur. I learned about them from Maggie Litz, the historian at the Winterthur Museum. Near what had once been the Winterthur train station (the du Ponts were wealthy enough to afford their own station) are a half-dozen old *C. crus-galli* standing in a line. Since they have not been maintained as a hedge for at least 150 years they have reverted to a simple formation of trees. There is no way of knowing whether they were ever coppiced or laid or whether they started life as a quick-set hedge. Although no mention of hawthorn hedging at Winterthur has been discovered in the farm account books, we can assume that the estate was laced with such living fences, and later with *Maclura pomifera.* The proof of the latter can be found along the eastern boundary of Winterthur, where, running for more than a third of a mile, is a low, dense Osage orange hedge that was planted around the turn of the century to replace a hedge that was probably planted in the mid-1800s.

There was so much demand for hedges that Delaware alone supported two private corporations, the New Castle County Hedge Company and the Delaware Hedge Company. These contractors used an implement operated by hand and foot that was patented in 1890 by Wesley Young of Dayton, Ohio. It worked via a pair of hinged heads that prevented soil from falling back in after a hole was dug. Before this invention, the patent application maintained, a planter inserted a spade in the earth at an angle, pried it open, and tossed in a seedling, hoping that when the ground closed around it the plant would be standing straight enough to grow. Apparently a lot of seedlings got bent over during the process and died. The device also sported blades at the bottom of the heads so that the planter could clip the roots and fit the seedling into the hole. Young patented several other hedging devices. These included a tool for bending over hedge plants, apparently a way of laying very young trees without cutting their trunks. How the planter prevented them from springing back afterward is not explained in the patent. But maybe to solve this problem Young later invented a contraption to accomplish what hedge layers did with stakes, woven branches and artful pruning: a four-wire lattice strung between posts that supported the saplings at the proper angle.[28]

The vast hawthorn hedges planted in the East began to decline as soon as farmers turned to *Maclura* and then metal. The decline was hastened by insect pests such as the apple-boring beetle. In addition, the increased use of coal for heating meant that large quantities of timber were no longer needed for fuel and could be used for fencing. As barbed wire and steel webbing became cheaper, farmers on the Plains abandoned the Osage orange, as well, tired of seeing their hard work ruined by prairie fires. Another of their complaints was that hedges created rigid field boundaries that are more difficult to alter than wire strung between posts. Just as in Europe after World War II, the need to consolidate fields became more pressing as farms and the machines used to work them grew bigger. The average American farm in 1860 was two hundred acres; in 2012 it had grown to more than four hundred. The first John Deere tractor had a 25-horsepower engine; the most ferocious of the modern John Deeres is 560.

Americans finally decided they would rather throw up a short-term mechanical barrier than spend time planting, growing, and tending a fence that would last for decades and provide critical elements of successful agriculture that wire cannot supply, such as protection from the wind, prevention of soil erosion, and flood control. In some parts of the United States there is no longer a demand for farm fencing at all. Vast fields of grain, soybeans, corn, and cotton sprawl unfenced because fences get in the way of gargantuan tractors and harvesters. As family farms become absorbed into corporate farms, one of the many things lost is the memory that hedges made from hawthorn and then Osage orange were responsible for the rise of American agriculture and the colonization of the West.

But one culture's material gain is often another culture's loss. As much as any other factor, fencing on the prairie allowed white settlers to steal land from Indians and deprive them of the bison that were at the core of their world. Because the U.S. military found the guerrilla style of hit-and-run warfare practiced by the Plains tribes hard to overcome, the only way to subjugate them was to fence them out. Although barbed wire was not *intended* to control people, like the hedges of England and Ireland that were being used to deny the underclasses access to common land during this same era, the outcome was the same. Wire strung across the prairies in the nineteenth century in order to separate cattle and sheep from crops also altered the migration patterns of the bison, and closed off land on which Indians once hunted at will. Faced with the loss of territory, coupled with the wanton slaughter of the herds and the introduction of European diseases, Indians finally succumbed and were confined to reservations.

But although the hawthorn played a role in the destruction of traditional American Indian cultures, the tree also offered them life.

SIX

The Return of the Native

> A healthy life is more than medicine. A healthy life is built from the inside and includes prevention of illness. To the Indians, food and medicine are largely the same thing.
>
> —ALMA HOGAN SNELL

The red sandstone foothills of the Bighorn Mountains descend toward Canada onto a wrinkled prairie, studded with low rises and laced with coulees. It looks like a tough place to make a living. But for years, working here in the middle of nowhere, five hundred miles from the nearest delicatessen, was a woman filling her pail with exotic fruit from the tangle of bushy trees crowding the banks of a dawdling stream. Her white hair framed an inviting face with high cheekbones and dark brown eyes. Her skin was the color of a penny. Except for her jeans and jacket she could have stepped out of a scene set on this very spot ten thousand years ago, when wanderers from Asia made their way across these high plains and saw enough promise in the place to stay.

Until her death in 2008 at the age of eighty-five, Alma Hogan Snell had hunted wild plants here on the Crow Indian Reservation in southeastern Montana for much of her life. She learned about Crow botany from her maternal grandmother, Pretty Shield, an important medicine woman. Her father was a shaman, Goes Ahead, and one of the four Crow scouts who urged George Armstrong Custer to wait for reinforcements before attacking the huge village of Sioux camped on the Little Big Horn River that famous June day in 1876. Pretty Shield learned from Strikes-with-an-Ax, her aunt. And so the lore was passed down, from woman to woman, across scores of generations. At its core was the idea that everything good has its time, Alma explained, and you should eat a little of every good thing. She lectured extensively about Crow botany, speaking in English, sign language, and Absarokee, the tongue still spoken by a third of the fourteen thousand natives living on the reservation or in nearby towns. She always stressed that "if nature could only speak right out to us, she'd say, 'This is what this contains, and you need it in your body.' She ripens what we need just in time for us to eat it."[1]

In June, for example, the *ehe* (pronounced ay-hay) is ready. *Ehe,* the Crow word for wild turnip, *Pediomelum esculentum,* is not related to the garden turnip but is a member of the pea family. At its best it tastes like a combination of uncooked green beans and corn, with notes of unroasted peanut. It is a perennial forb, hairy, with five fleshy leaves growing at the end of gangly stalks, and purple-blue flowers sprouting from the ends of their own stalks. The edible part of ehe grows as a tuber the size of a chicken egg on the root. For the Plains tribes this plant, also called bread root, has been a staple, high in minerals and carbohydrates, for generations. Ehe can be eaten raw, cooked in stews, and added to porridge as a thickener. But it is mainly used as a flour, dried and ground into a fine meal that can used to make a simple griddle bread or combined with wheat flour to add an exotic note to fry bread.

Deciding precisely when the ehe is at its best is a challenge. Harvest it too early and the tuber will be emaciated. Wait too long and it will be tough and gnarly and taste like wood. (While you're harvesting, make sure that you're not picking lupine, which looks a lot like ehe and can flourish beside it in the same patch of ground. An alkaloid called lupinine is a neurological toxin that can cause hallucinations and seizures in people and death in foraging animals.)

After the ehe has been harvested it's time for prairie onion, *Allium textile.* When the rough brown skin is stripped from the root it reveals a white bulb tasting of garlic that injects a robust pungency into stews. In July the wild carrots, *Perideridia gairdneri,* will be also be maturing. These are added to stews, mashed into puddings, or tossed in salad with watercress, cattail stalks and seeds, white yucca flowers, the pink flowers of the wild rose, and maybe roasted turnip-bread croutons, all dressed with oil and juice squeezed from prairie grapes.

As the summer winds down people pick fruit. From the creek bottoms they take wild plums, serviceberries, currants, grapes, rosehips, and buffalo berries, as well as mint and pennyroyal to be dried for tea. Up on the bluffs they collect juneberries, elderberries, gooseberries, prickly pears from the ubiquitous prairie cactus, red berries from the little pincushion cactus, and kinnikinick, which the Crow call *Obeezia,* or bear berry. Some of this food will be eaten raw, but most of it will end up in jams, syrups, pies, and mason jars.

Some of the plants won't go into the pantry; they'll go in the medicine cabinet. Chipmunk tail, or yarrow, is used to treat sunburn, cuts, and stings. Long before *Echinacea* became the rage as a cure for the common cold (despite the results of clinical trials indicating that the herb is useless against cold viruses), the Crow dosed themselves with it to fortify their defenses against infections and respiratory ills. Pretty Shield taught her granddaughter how to make a tincture of *Echinacea*—what the old Crow called black root or snake root—that is used to ease the pain of toothaches and treat canker sores.

After the first frosts of October have colored the leaves and bleached the fescue, the final harvest of the year is offered by two species of tree whose fruit is at its sweetest because of the cold nights. Flourishing together in dense hedgerows that line the coulees, these are chokecherry and hawthorn, called *beelee chee shah yeah* in Absarokee. In wet years the hawthorns sag with fat berries that are so dark a purple they are almost black. How dense are these hawthorn thickets? Tim McCleary, an anthropology professor at Little Bighorn College, conducted a timber survey of the Crow Reservation in his capacity as archaeologist for the tribe's Historic Preservation Office. Tasked with ensuring that certain cultural and historical sites would not be disturbed by logging operations, during the summers of 2001 and 2004 he worked along Thompson Creek in the Wolf Mountains fifteen miles east of the Custer Battlefield. McCleary, who was a good friend and associate of Alma Snell, told me that hawthorn surrounds the mountains at the base of the foothills and is "incredibly difficult" in some places. "I often searched for breaks that had been made by cattle and then I noticed that the older cows got through the thickets by walking backwards. I tried it and found it much more successful then ducking or crawling head-on."[2] When they were little girls Alma and her sisters rode their horses bareback to places such as Akbilitchishée Aashkaate, or Hawthorn Bushes Creek, where all kinds of fruit grows. There they gorged on berries till they couldn't eat any more, then filled their baskets and rode back to the house they shared with Pretty Shield up in the foothills. When Pretty Shield foraged for fruit she had to face two hazards her granddaughters were spared: grizzly bears, and men from other tribes who kidnapped Crow women and kept them as slaves. Before the "white bears" abandoned the Plains for the mountains after the bison herds were exterminated, they prized the berry thickets in autumn, and they didn't like to share them with people.

In the old days the women mashed chokecherries—pits, skins, and all—with rocks and shaped them into patties that were dried in the sun and put by for winter. An even finer mash is still mixed with chokecherries, dried pulverized bison meat, and the tallow from bison kidneys to make pemmican. Although the herds are no longer so numerous that they replace the tans and greens of the prairies with their own dark umber, the Crow Nation grazes fifteen hundred head in the Bighorn Mountains, the largest bison herd in Indian country.

But the principal value of hawthorn for the Crow has always been medicinal. The tree was thought to supply an agent that broke apart blood clots in the circulatory system, relieved chest pains, kept the heartbeat regular, and was a general tonic for strengthening heart muscles and pushing blood. *Crataegus* was considered such a vital part of the traditional Crow pharmacopoeia that before the women picked anything from a hawthorn tree they would offer thanks to the Unseen Force for loving them so much that it provided them with this bounty. And they would promise to make the tree a pair of moccasins. This was a way of paying the hawthorn back—shoes for its feet, its roots.

Alma Snell told a story about her son's mother-in-law, who needed surgery to repair damage to her heart caused by disease. The doctors informed her children that because her heart was so weak she might not survive the scalpel, but surgery was the only thing that would keep her alive. Alma recommended that she take hawthorn. For two weeks, with her doctor's permission, the woman took large doses of *Crataegus* gelcaps bought over the counter from a health-food store. "When she went into surgery," Alma wrote, "nobody knew what would happen. They didn't know whether the hawthorn would work or not. But she survived the procedure and afterward the doctor said, 'It was amazing to me. The muscles of her heart were so strong. Nothing at all like the weakness we found just a few weeks before.'"[3]

On a trip to Washington, D.C., in the mid-1990s, when she served on the advisory council for the design of the National Museum of the American Indian, Alma was approached at a meeting by a man who told her that multiple sclerosis had ended his son's career as a professional musician and asked whether Alma could recommend something to help him "I instructed him to take hawthorn at the highest strength sold and to follow the directions on the bottle. And so he did. The next month I was in Washington again. This same man rushed over to me and said, 'I want my whole family to meet you. I want to take them to Montana to meet you because of what you have done for my son. He started taking that hawthorn, and now he's playing again. A month later, and he's back on the job.'"[4]

If hawthorn can be bought in the supermarket, why would people work so hard to harvest wild plants? One answer is that the price is right. During tough economic times people turn to their gardens, fishing poles, and hunting rifles. And growing numbers of foragers, inspired by classics such as British author Richard Mabey's *Food for Free,* are using tender spring hawthorn leaves in salads. Recognizing the trend, the *The Guardian* published an article encouraging its readers to use this early foliage to add a nutty taste to potato salad and cheese sandwiches.[5]

But economics were not why Alma sought the wild hawthorn. Nor did she forage because she lacked experience in the culinary arts—she was employed her entire adult life in the food business, running her own café and working for hospital food services and restaurants. The real reason to harvest wild plants, Alma claimed, was that "there is value in our Crow foods and medical substances that can ward off many sicknesses. If we practice our ancestors' ways of preparing food and living healthfully, we may be able to prevent a lot of ills."[6] (While a couple of small, boutique suppliers—such as Wild Pantry in Tennessee, which offers a pound of the dried fruit of yellow hawthorn, *C. flava,* for forty-five dollars—sell freshly harvested wild plants, you've got to go out and get things such as ehe and prairie onions for yourself.)

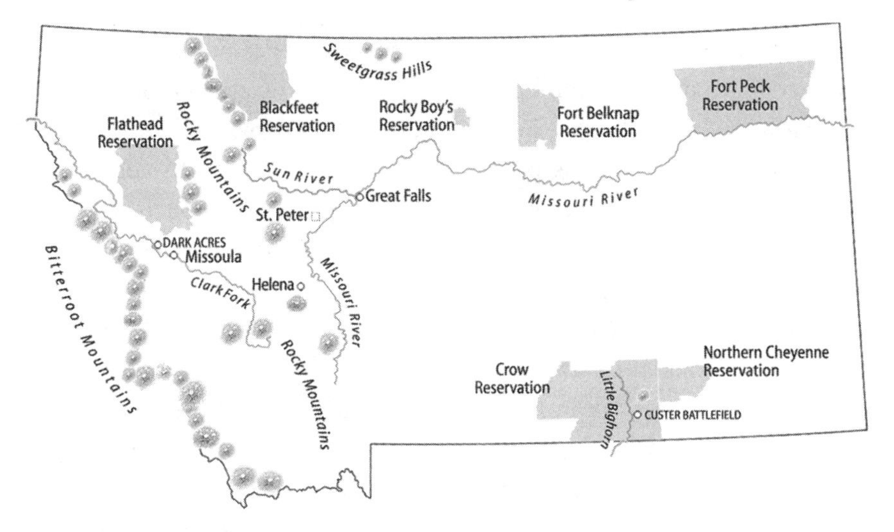

The seven Indian reservations within the borders of Montana (map by the author)

You don't have to look very hard around Crow Nation to see what sort of ills Alma was talking about. Health problems caused by obesity have become a serious issue on reservations. One cause for its predominance among the Crow may have to do with the genetics of these traditional hunter-gatherers: the Crow have become much more sedentary and share little in common with their nomadic warrior ancestors, whose lives were extremely active physically only six generations ago.

To address the poor state of reservation health, tribal leaders are encouraging their people to put down the pizza and return to the unprocessed diet that sustained them for thousands of years. The bodies of Plains tribal people developed in response to a diet that was low in carbohydrates, fats, and sugars, and high in fiber and protein. One theory about obesity holds that Plains Indian metabolism is driven by a "thrifty gene" selected by the forces of evolution because it orders the immediate conversion of surplus food into body fat. This mechanism would allow people to quickly stockpile energy during times of plenty for times when food runs out. In terms of the survival of the species, females with an adequate supply of body fat stand a better chance of getting pregnant and producing healthy offspring than thinner females. But today, bodies that evolved in scarcity are immersed in relative bounty, and relentlessly store energy for a famine that never arrives.[7]

Despite its intuitive logic, the "thrifty gene" theory has come under repeated attack since it was proposed in 1962. One reason is that although farmers indeed suffer from hunger and starvation when their crops fail (witness the Potato Famine in Ireland), we have no conclusive evidence that famine was a serious problem for hunter-gatherers. In addition, although many of the twenty to twenty-five thousand protein-coding genes in the human genome have a recognizable function, geneticists have not been able to positively identify any of them as "thrifty."[8]

Another theory, concerning the "drifty gene," centers on the evolutionary force of genetic drift, or the random tendency of certain traits to become more prevalent in a population (I had more kids than you did so my genes, good and bad, got passed on more frequently). Some have argued that a rare obese body type began appearing more frequently two million years ago, when humans developed enough technological and organizational skills to routinely avoid serving as the dinners of animals.[9]

Whatever the reason for the increased girth of American Indians, the result has been a plague of hypertension, heart disease, and diabetes. More than 16 percent of tribal adults were treated in 2012 by the Indian Health Service for diabetes, almost twice the national average. About 12 percent of Crow have the disease, the highest percentage among Montana's seven reservations and twice as high as the rate for the state as a whole. Alma Snell told her people that despite its challenges, following the traditional paths was the only way the Crow could restore themselves to wellness. "I do believe that nature provides at each step along the way," she said. "We never had all this year-round choice of food as we do now. We had each thing in its season. We walked with nature. Nature provided, and we were healthy."[10]

The tribes of the Northern Plains shared much more than their nomadic way of life. For one thing, the economies of the Crow and their traditional enemies—the Cheyenne, Sioux, and Blackfeet—were almost identical. Bison, game animals, birds, and plants, not the least of which was the hawthorn, were at the center of their cuisine, technology, and pharmacology.

The Sioux used the leaves, berries, and flowers of the hawthorn for food and pharmacy, just as the Crow did. And some tribes ascribed more obscure and untranslatable powers to the plant. An important force in the cosmology of Arapaho and other Plains tribes, for example, was thunder and the manifestation of this force, the Thunderbird. Hawthorn berries are called *baa-ni-bia,* "thunder berries," in Arapaho and were believed to be a gift to the people from this power.[11]

In the northwest quarter of Montana, opposite the Crow Nation, lies the Blackfeet Indian Reservation. Bordering Glacier Park and Canada, much of it, like its Crow counterpart, is prairie. But the traditional territory the Blackfeet ruled encompassed a stretch of the Rocky Mountains, and the wild plants they harvested included species such as the glacier lily and the camas, whose bulbous roots are baked in order to convert an inedible polysaccharide called inulin, which is their main component, into the fructose form of sugar. One of the medicines Blackfeet women forage for is alumroot, which is used to dry wounds and is brewed into a tea to combat inflammation. Another is the juniper berry, used to treat kidney problems. Ripe haws from the fireberry hawthorn, *C. chrysocarpa,* called *i'kaasi'miin* in Siksika, the Blackfeet language, are harvested for food to some extent, but were used traditionally as a laxative and a heart tonic. The berries were immersed in boiling water, which extracted a tea. The tree's hard wood was also useful; it made a good digging stick to unearth camas bulbs.[12]

Rosalyn LaPier, a historian at the University of Montana, learned about the two hundred or so plants regularly used by traditional Blackfeet in Montana and Alberta from her grandmother Annie Mad Plume Wall, who was a member of the Never Laughs Band that settled after the Indian wars on Badger Creek, which flows east from Glacier Park. Wall, who foraged for food and medicine almost every year of her life from May to October until her death in 2009 at the age of ninety-five, was taught by *her* grandmother Not Real Beaver Woman, who was taught by her mother, Big Mountain Lion Woman, and so on. Because Blackfeet territory gets more rainfall than that of the Crow there are more varieties of berries available for food and medicine: at least eighteen different species. Arguably the tastiest of these is the huckleberry, which is picked in late summer. Although Wall lectured that huckleberries should be used only to improve eyesight and strengthen the cardiovascular system, they were and still are eaten raw on the reservation and added to baked desserts such as pies. In matters of taste hawthorn fruit, which contains less than 2 percent fat, cannot compete with the huckleberry. LaPier told me that during the dark days after her people were herded onto the reservation the hawthorn berries were used as "emergency food," prized because they stay on the branches all winter (unless birds get to them before the people do).[13]

If all of North America's hawthorn species were combined the range of the tree could be said to stretch from ocean to ocean and from Nicaragua to the Arctic Circle. Consequently, various *Crataegus* parts were on the menus and in the medicine cabinets of almost every one of the continent's indigenous cultures. Based on its nearly universal use, hawthorn was arguably one of the most important plants in pre-Columbian America. Along the west coast of Canada and the Pacific Northwest tribes such as the Colville and the Kootenay ate the berries with salmon roe and bear meat. Sometimes they baked the berries on slabs of wood laid near the campfire, creating a food with the consistency of raisins, or pounded them, like pemmican, with dried salmon. The berries were also mashed and formed into thin, hard cakes, which were dried and dipped in soup.[14]

One hawthorn species, *C. mexicana,* grows in clearings in the pine and oak forests of Mexico. The Aztecs ate its yellow berries, but it was primarily used as a diuretic, a fortification for the muscles, and a tonic for the treatment of high blood pressure. The roots of *C. mexicana,* which was called *tejocote* or *manzanita* by the Spanish, and *texocotl* by the Aztecs, were boiled in water until much of the water had evaporated, leaving behind a strong tea.

The Fox Indians of the Midwest made an infusion of twigs and root bark from the pear hawthorn, *C. calpodendrum,* to treat bladder problems. The Potawatomi Indians living in the upper Mississippi River region used the fruit to calm upset stomachs. The Kwakiutl people on Vancouver Island chewed the leaves to make a poultice that was applied to swellings. The Cherokee of the southeastern United States used the berries of the littlehip hawthorn, *C. spathulata,* as a heart medicine and to calm muscle spasms.

It is not known precisely why so many different cultures in North America and around the world began the practice of using hawthorn for medicinal purposes. As with most wild plants harvested for food and medication, trial and error was probably involved, establishing methods of preparation for each plant, as well as servings and doses. If something tasted good and didn't make people sick, or tasted bad but made them better, people would make it part of their diet or include it on their list of medicines.

Paleolithic newcomers to North America got ideas about good and bad plants by watching animals. Like dogs, which eat grass when they are constipated, many animals self-medicate with plants, soils, and insects, a phenomenon called zoopharmacognosy. (Another example is pregnant animals that eat blue or black cohosh, *Actaea racemosa,* a leafy herb that induces birth by stimulating contractions in the cervix.) The Blackfeet believe that bears, which are omnivorous, taught humans about edible plants. Watching bears eat hawthorn berries would show people that the fruit probably would not kill them. When the bear ate the leaves people may have experimented with that part of the tree, as well, and discovered an interesting new taste. And after hawthorn became part of their diet some people might have noticed that they felt better and had more energy, and may even have made a connection between their ingestion of hawthorn and a stronger heart. When hunter-gatherers traded with their neighbors, information as well as material things were passed between them, and the good news about hawthorn might have spread. (The extent of the pre-Columbian trade network in North America is illustrated by a pair of gorgets made from large sea shells that were found in a cave in the Sweet Grass Hills near the Blackfeet Reservation. Carved to resemble human faces, they were passed from hand to hand six hundred years ago all the way to Montana from the Gulf Coast.)[15]

Traditional aboriginal people in America did not approach medicine in the prosaic way Europeans did. Among the people of the Northern Plains, ceremony and prayer were attached to every daily act. An unending communion with forces of the unseen world was the same thing as consciousness. Since medicinal plants were regarded as animate beings, the Plains Indians would speak to them, reminding the plant of its purpose, thanking it for its help, and invoking higher powers to ensure that it would restore the balance in their bodies.

In addition to its value as food and medicine, the hawthorn had other uses. Most native peoples in North America fashioned weapons and implements from the wood because of its fine tight grain, resistance to rot, and exceptional hardness. Like the Blackfeet, the Salish of Montana valued the wood of *C. douglasii* for use as camas sticks used to unearth these and other roots. These digging tools were two feet long, curved like a saber, and tipped with a sharp point hardened even further in fire. A crosspiece of elkhorn strapped to the other end served as a handle. In a nineteenth-century study, a botanist named Charles Geyer noted that the sticks were used with "astonishing dexterity."[16]

Eastern Woodland tribes such as the Seneca, the Mohawk, and the Iroquois valued hawthorn for the speed with which it snaps back to its original shape after being flexed. This quality, generated by the dense, interlocking fiber in the wood, resulted in a bow that demands a considerable amount of muscle to draw but sends an arrow to its target with a higher velocity than that produced by softer, more porous woods, such as the chokecherry (from which many of the short, three-foot bows that rained clouds of arrows down on Custer were made). Sometimes weapons made from hawthorn were asymmetrical, crafted from one of the snaky limbs of the tree. The limb would be rubbed with beeswax or fat so it would dry more slowly and thus minimize the splitting and rupturing that can occur when wood cures. Then the bark would be peeled away, and the shape of the bow would be refined by heating it over a fire and working it with the hands. Finally, it was rasped and sanded with stone tools—and later iron tools—to make its satiny smooth skin fit neatly into a man's hand.

If the warrior preferred a symmetrical bow the process required more time and better tools. Michael Bittl is a professional bowyer who lives on an old farm in Germany's Black Forest with his wife, daughter, and a menagerie of animals. Sought after by historians, museums, and universities because of his knowledge of old and ancient bows and arrows, he has crafted several hawthorn bows employing methods used by the Woodland tribes. He begins by hiking into the forest in late winter, when the sap in the fibers is at its lowest volume, and therefore the wood is as dry as possible, and looks for a hawthorn trunk that is as straight as hawthorn trunks ever grow and the right diameter—between four and six inches is a good, workable girth. This he takes back to his shop, where he cuts a length of trunk four feet long.[17]

Then comes the challenging task of splitting the trunk into two staves. Because hawthorn's fibers are tightly connected to each other—unlike the fibers of, say, ponderosa pine—considerable care and effort must go into halving the trunk. Indians used stone wedges and hammers. Bittl uses steel to start the split and finishes the job with a handsaw. The staves are rubbed with beeswax, tightly clamped against lengths of straight milled timber, and left alone for several weeks to dry. (Indians put the staves up on the ceilings of their longhouses, where heat and smoke tended to collect.) Once the wood is cured, he removes the bark, shapes it using heat, and sands it. He might then reinforce it by wrapping it with sinew. Finally, he chisels a groove around each end to serve as channels for the bowstring, which Indians made of hemp or sinew, and sometimes the inner bark of hawthorn braided into twine.

In the same compulsion to economize that compelled the Plains Indians to find a use for every single part of a bison—as a child Pretty Shield played with a ball made from the lining of the heart, stuffed with prairie grass and leaves—aboriginal people put the thorns of *Crataegus* to work, as well as the rest of the plant. West Coast tribes used them as fishhooks, and all North American tribes with access to them used them as points for lancing boils and removing splinters. The Iroquois, whose spiritual world included a belief in witchcraft, fashioned a sort of hawthorn voodoo doll to put a hex on an enemy, making him "break out like cancer," according to one oral history. This was most commonly the root of Quebec hawthorn, *C. submollis,* which would be carved to resemble the victim, or it could be another material, like corn husks, which would be fashioned into a doll and filled with thorns or pierced with them. Conversely, to ward off such a spell the victim countered hawthorn malevolence by drinking hawthorn tea.[18] Among some Pacific Indians, *Crataegus* thorns are believed to be toxic and are decocted to make a poison. This erroneous belief probably has its source in the fact that puncture wounds from something as sharp as one of these thorns often do not bleed enough to cleanse the flesh of bacteria or fungi, which have developed a symbiotic relationship with the thorns of some species, an innovative act of evolution I'll discuss in Chapter 10.

The Northern Plains Indians laced a thorn onto a stick or piece of bone to make a sewing awl. The awl pierced the animal skin being made into a garment, and a length of string made of sinew or hemp was inserted into the hole and pulled through. Experimenting with a nasty two-and-a-half-inch thorn from a branch of fleshy hawthorn, *C. succulenta,* that I harvested in central Montana, I easily poked a hole in my soft leather briefcase. Such a sewing awl was an important prop in the final act of Custer's Last Stand. When his naked, bloated body was discovered two days after he had been killed on what is now the Crow Reservation, it was discovered that Custer had been spared the ritual mutilation inflicted on the bodies of most of his men, which was largely carried out by the women and children of the village. The only visible damage was the bullet hole in the front of his left temple, and the bullet hole in his chest above his heart. But according to Kate Big Head, a Cheyenne who witnessed the battle firsthand, there were invisible wounds, as well. She said that two women from her tribe bent down to remind Custer that when he smoked a peace pipe with the Cheyenne's chiefs they told him that if he ever broke his peace promise the "Everywhere Spirit" would see to it that he was killed.

Then, to improve his hearing, they drove a sewing awl into each of his ears.[19]

When the people who would become American Indians crossed into North America from Asia they brought with them a library of knowledge about botany. While they learned how to use many new species they found in the new world for food and medicine by trial and error, some of the familiar plants they came across—such as the hawthorns—were very much like the ones they had left behind.

SEVEN

The Tree of Heroes

If agriculture provides neither better diet, nor greater dietary reliability, nor greater ease, but conversely appears to provide a poorer diet, less reliability, with great labor costs, why does anyone become a farmer?

—MARK NATHAN COHEN

For a decade the People's Republic of China was embroiled in a bizarre ideological civil war called the Great Proletarian Cultural Revolution. Raging from 1966 to 1976, it was the result of a series of massive upheavals beginning with the second Sino-Japanese War, followed in 1949 by revolution, when Communist troops under Mao Zedong drove the Nationalists under Chiang Kai-shek from the mainland to Taiwan. Then, in 1958, Mao initiated the Great Leap Forward, an effort to organize millions of people into collectives for the production of commodities. Many of these units would be assigned to fabricate a single item, steel, in order to push China's largely agrarian society into the industrialized world. (China's obsession with steel continues today, and its manufacture is the major source of the country's dangerously polluted environment.) But the Great Leap Forward was a great stumble backward because the metal, produced by untrained farmers, was of such poor quality it was useless. Food shortages plunged China into widespread famine, and in 1960 Mao was forced to resign as head of state.[1]

He attempted to regain power in 1966, when he campaigned for the suppression of what he declared were "bourgeois" elements infiltrating government and society, intent on crushing communism and restoring capitalism. His message was aimed at young people, who responded with the idealistic zeal of the naive. Soon, gangs calling themselves Red Guards began wandering the cities, waving Chairman Mao's collection of political aphorisms, the *Little Red Book,* persecuting and attacking people who had the "wrong" politics or came from the "wrong" class. Educated city-dwellers were sent into the countryside to learn from the peasantry, the class that Mao was from and for whom he claimed to have waged the revolution. In their eagerness to obey Mao's dictum "Smash the Four Olds"—old habits, old ideas, old customs, and old culture—the Red Guards burned books, defaced statues of Buddha, destroyed churches and temples, and ruined objects they looted from museums. Tibetan monks were forced at gunpoint to destroy their monasteries. The arts became dominated by awkward, fervent sloganeering, socialist realism, and the tenets of Marxist-Leninist thought as painters, playwrights, and musicians deemed old school or anti-socialist were harassed, blacklisted, and killed. (The most popular song of the era was a propaganda piece called "The East Is Red.") Political shibboleths were printed on everything from cigarette packages to bus tickets. Even personal relationships came under scrutiny. The authority of parents was questioned for the first time. Young people were asked to put off marriage and were encouraged to marry only someone with the "right" background—that is, a member of the working class, a soldier, or a peasant. Pitting class against class, the violent turmoil resulted in an unknown number of deaths, possibly several million. Meanwhile, the economy ground to a standstill.[2]

In 2007, an anonymous blogger using the alias Ai Mi, tentatively identified as a Chinese woman in her sixties living in Florida, posted a novel on her blog called *Hawthorn Tree Forever* that was set against this tumultuous backdrop. The blog was blocked by the Chinese state, but a government-sanctioned publishing house picked it up and released it as a book titled *Under the Hawthorn Tree,* which became a huge success, selling millions of copies. A film version with the same title was released in mainland China in 2010 and became an instant blockbuster. (It was directed by Zhang Yimou, who also directed such international hits as *House of Flying Daggers* and *Raise the Red Lantern.*)

In the book we are introduced to an innocent, hard-working high school senior named Zhang Jing Qiu. Her father is a landlord—in Mao's view the most despicable segment of society—who has been sent to a labor camp in the countryside for "re-education." Jing Qiu shares a shack in the city of Yichang in central China on the Yangtze River with her mother, a school teacher, and her younger siblings. Like most Chinese of the period (except for ranking members of the military and the Communist Party), they live in relentless poverty.[3]

Because of her skill as a writer, Jing Qiu is one of four students selected to visit West Village, a rural backwater located in the dramatically picturesque Three Gorges Area. Their assignment is to interview the peasants and compile a history for their school to replace the old, politically incorrect textbooks that were full of feudalism, capitalism, and "revisionism." Trudging up a mountain ridge in the cold air of late winter, they follow the mayor of the village, who promises to let them rest when they reach the hawthorn tree. Jing Qiu is eager not just to get off her tired feet but to see a hawthorn. She expects something magical, for she once learned a Russian song called "The Hawthorn Tree" that told the story of two young men who were in love with the same young maiden. Because the maiden liked them both and could not decide which one to choose she went to a hawthorn tree to ask for advice.

> Oh! Sweet hawthorn tree, white buds on your branches,
> Ah! Dear hawthorn tree, why so troubled?
> Which is the bravest? Which is lovelier?
> Oh, I beg you, hawthorn tree, tell me which one.

But when the visitors reach the hawthorn, it is hardly magical, much less infused with the romantic aura of the song. Twenty feet high, gnarled, and bristling with spikes, it has not yet sprouted a single leaf or flower. Jing Qiu asks the mayor what color the blossoms will be when it blooms and is surprised when he answers "red." Take out your notebooks, he tells them, because this is the first story for your book: During the war against the Japanese, many brave Chinese soldiers from West Village were marched to this spot by the invaders. And here they were executed. Their blood watered the tree, which had always produced white blooms before the war. In the novel Jing Qiu decides to title this chapter of her group's village history "The Red Hawthorn Tree," though in the movie version the hawthorn is called the Tree of Heroes. As they head back to the village she looks behind her and sees in the distance a young man in a white shirt standing under the tree.

It has been arranged that Jin Qiu will stay with the mayor and his family while she works on the textbook, and on her first night there she meets a friend of the family, a twenty-six-year-old soldier named Sun Jianxin with whom she is instantly smitten. As they chat and he flirts she asks him about the red hawthorn tree, and he tells her that it is not scientifically possible for blood absorbed by the roots to change white blossoms into red. But he also notes that it is not her job to report the truth—merely to record what people say. Since she will be returning to the city at the end of April he offers to write to let her know when the hawthorn is blooming so she can come back and see it.

Despite the fact that Jianxin is from a "correct" background and Jing Qiu's is incorrect, they start seeing each other. On an outing they find themselves after dark near the Tree of Heroes, and he asks her if she'd like to go look at it. But she's afraid: The day I came to West Village, she says, I saw a man in a white shirt standing under the tree. Was that you? He says no and to tease her says that a man in a white shirt is standing there now. Is it the ghost of a Japanese soldier? Terrified, she starts to run but he pulls her into his arms and kisses her.

So begins Jing Qiu's education in a love that would never be consummated. After she returns to the city Jianxin continues to pursue her, but she is afraid of her feelings and worried that her one act of intimacy could ruin her life and make her family's difficulties even worse. (Her mother has recently been denounced as a "counterrevolutionary.") One day he secretly leaves a vase full of red hawthorn blossoms outside the door of her shack. She puts them next to her bed. And so it goes. She wants to be with him, but is afraid of what people will think, so they meet from time to time in an isolated place where no one will recognize her.

Back in the city her days as a high school student are filled with classes and sports and odd jobs to help out her family. Hovering over her future is the likelihood that after she graduates she will be sent to work in a forest. This policy of forcing "educated youths"—that is, high school graduates—to labor beside peasants was called the Down to the Countryside Movement. Jing Qiu has heard stories about how difficult and boring life is for people after they are sent down. (Mao's *Little Red Book* does not quote Marx's comment in the *Communist Manifesto* about the "idiocy of rural life.") But she is saved from this fate when a new state policy allows the children of teachers to take over their parents' positions when the parents retire. Jing Qiu's mother is willing to step down, and Jing Qiu is ready to take her place.

Jing Qiu's mother finally finds out about her daughter's relationship with Jianxin and extracts a promise from him to stay away from Jing Qiu for thirteen months, until her appointment as a teacher is finalized. But they spend an intimate night together, though they stop short of the sexual act. Later, after a long separation, she learns that he has leukemia—blamed on radiation from the U.S. atomic bombs dropped on Japan—and is on his deathbed. When she goes to him in the hospital she learns that he has been hanging on so he can say good-bye to her. After his death Jianxin is cremated and his ashes are buried under the hawthorn tree.

The novel belongs to a genre of fiction known in China as "scar literature," which is set during the Cultural Revolution. While these works are mainly popular among people who survived those wrenching years, *Under the Hawthorn Tree*—which is reputedly based on diaries kept by a friend of the author—appealed to younger readers as well. The red-blossomed hawthorn symbolizes enduring love, impervious to the debilitating state-sponsored dogma that condemned romantic love as an archaic, antisocial indulgence. But it also represents the ancient legacy of Chinese culture, a legacy that not even the profound dislocations of the past sixty years has destroyed. Feudalism gave way to socialism, a Communist government encouraged state-sponsored capitalism to flower, fifteen hundred villages and towns in the Three Gorges Area were flooded by water backed up behind the largest dam in the world, and the government decreed that couples could have only one child (Zhang Yimou, the director of *Under the Hawthorn Tree,* was fined $1.3 million after it was discovered in 2013 that he and his wife had three children), but vestiges of the culture endure.[4] One of those vestiges is belief in the efficacy of the hawthorn. Although the state tried unsuccessfully to suppress Buddhism, Christianity, Islam, and Daoism, and Mao gave permission to one of his subordinates to dig up the grave of Confucius, he did not attempt to ban traditional Chinese medicine, even though it qualified as one of the "Four Olds." This is particularly surprising since the practice is intrinsically bound up with Daoism, which is more a philosophy than a faith. Since the hawthorn has long been used in Chinese medicine and cuisine it is no coincidence that "Ai Mi" chose it as her symbol.

Nine thousand years ago people living in a vibrant little community called Jiahu in China's Yellow River basin kept dogs, grew rice and foxtail millet, and may have domesticated pigs. They fired pottery, played musical instruments, ceremonially buried their dead, hunted with bows and arrows and harpoons, scratched signs on shells and bones that are thought to be a form of early writing, and communed with the Unseen Force through the medium of a shaman. Jiahu represented one of humankind's most dramatic breaks with the past, a true great leap forward, which had profound outcomes, both positive and negative. Instead of relying solely on hunting and gathering and the nomadic life these occupations required, the Neolithic Revolution ushered in an era of year-round settlements where food could be stockpiled and protected. As farming and the domestication of animals gradually replaced hunting and gathering, humankind's population boomed from the Nile to the Yellow River.[5]

Jiahu was discovered in 1962 by the director of the Wuyand County Museum. But productive excavations of the site were not undertaken for another twenty years, either because of the disruptions of the Cultural Revolution or because its significance was overlooked until archaeologists realized that the village was part of a network of related, but, as it turned out, younger and less important settlements excavated in the same area. Occupying almost fourteen acres, of which fewer than 5 percent have been brought to light, the oval-shaped ruin of Jiahu lies under several feet of earth on the slope of a mountain near what had been a swamp. To date, archaeologists have discovered the foundations of many small, one-room houses that were built partially below the surface of the ground, storage cellars, pottery kilns, graves, three-legged cooking stoves, weapons, and thousands of objects made of bone, pottery, stone, and other materials. Jiahu was continuously occupied for thirteen hundred years until it was destroyed by a flood. People returned to live at the site two thousand years ago.

The village was surrounded by a moat and organized into zones—residences, workshops, the cemetery. Some four hundred skeletons have been excavated, revealing that the people were of Mongolian stock and on average almost six feet tall, in stature similar to modern Chinese. Some Jiahuans, mostly men, suffered from arthritis caused by the stress of repetitive physical exertion. Other skeletons showed signs of anemia caused by an iron deficiency, evidence that as children they did not eat enough meat. Most villagers did not live past the age of forty. Although it is thought that their society was relatively egalitarian in practice, burial offerings indicate that some Jiahuans were more equal than others. One of the most compelling discoveries in a few of these graves were intact flutes made from the hollow wing bones of the red-crowned crane, a five-foot-tall bird whose beauty and theatrical mating dance form a recurring theme in Chinese art and literature. When researchers asked a professional flautist to play one of the instruments, it produced a resonant, haunting sound that had not been heard for at least seventy-seven hundred years. Archaeologists speculate that the graves containing these flutes were occupied by shamans who played them during attempts to contact the Unseen Force. Some of the skeletons were adorned with imported jade and turquoise jewelry, but millstones, awls, and other tools were also placed in these graves, evidence that even the more exalted members of society had to work. While scientists were excited about finding some of the world's oldest pottery at Jiahu, what was discovered inside the urns floored them.

The molecular archaeologist Patrick McGovern studies the past by analyzing the chemistry of ancient food substances in his role as the scientific director of the Biomolecular Archaeology Laboratory for Cuisine, Fermented Beverages, and Health at the University of Pennsylvania Museum of Archaeology and Anthropology in Philadelphia. In 1999 he traveled to the Institute of Archaeology in Zhengshou to examine pottery vessels unearthed at Jiahu. Up to eight inches tall, these elegant round jars have high necks with flaring rims and handles that resemble little ears. They were made by layering coils of clay, smoothing the surfaces, glazing them with a red slip, then firing them in a wood-fired kiln heated as high as 1,500 degrees Fahrenheit. When he peered into one of the vessels McGovern was surprised to see a reddish stain at the base of the jars rising up the sides. It dawned on him that he might be looking at the residue left by the world's earliest known alcoholic beverage.[6]

Using methanol and chloroform to extract material from sixteen stained pottery shards, then subjecting the extracts to a variety of chemical analyses, researchers discovered that the same compounds showed up on shard after shard. These included the chemical fingerprint of beeswax, which indicated the presence of honey. Chemicals called phytosterols pointed to rice as another ingredient. And the presence of tartaric acid told researchers that the third ingredient was a fruit. Because there are more than fifty species of wild grape native to China (none of which has ever been domesticated), it was assumed a grape was probably the mystery ingredient. But a subsequent discovery pointed to a second source of tartaric acid in the Jiahu vessels. A Chinese archaeologist did find seeds from wild grapes at the village site, but also seeds from another plant—hawthorn. "I suspect that both were added to the Jiahu beverage," McGovern wrote in his *Uncorking the Past: The Quest for Wine, Beer and Other Alcoholic Beverages,* "to add flavor and encourage fermentation."

Researchers concluded that the hawthorn fruit came from *C. cuneata* or *C. pinnatifida,* or possibly both. The first species, called cuneate hawthorn in the West and *shan li hung* in China, grows wild over much of the southern half of the country. More a bush than a tree, it rarely reaches heights greater than ten feet. It bears bright red or sometimes yellow haws a half-inch in diameter with green flesh inside. The flowers are usually intensely white with attractive red stamens, but some specimens produce deep rose blossoms. *C. cuneata* is cultivated for its fruit in China and Japan and is a favorite subject for bonsai artists. *C. pinnatifida,* the Chinese hawthorn, or *shan zha,* is taller, sometimes reaching twenty feet, and is different from most *Crataegus* species because it is usually not armed with thorns. Its native range is restricted to the northern part of eastern China. The luscious fruit, one of the best-tasting of all hawthorn species, ripens to a deep red and grows to be about one and a half inches in diameter. A valuable seedless variety has been developed for the food industry.[7] The skin of the fruit is flecked with tiny, lighter-colored corky pores called lenticels such as those found on apples and pears. These allow carbon dioxide to enter directly into the flesh and oxygen to exit. While its direct form of respiration gives the species a reproductive advantage by allowing it to grow big haws that birds and animals find irresistible, many diseases can also infect the fruit through these pores.

When McGovern and his fellow researchers were finished with their analysis they were fairly certain that the Jiahu beverage had been concocted from haws and wild grapes fermented with a wildflower honey mead and blended with a beer made from wild or cultivated rice. This Neolithic grog had an alcohol content of 9 to 10 percent, whereas most modern beers range from 3 to 7 percent, and grape wines from 8 to 14 percent. McGovern, who possesses the jovial demeanor, generous belly, and white hair and beard that would give him an edge on the competition in applying for the job of Macy's Santa Claus, decided that since he had figured out the ingredients in Jiahu's beverage of choice, he would make some. Re-creating an ancient alcoholic beverage was something he had done before, in the 1990s, when his team analyzed material taken from what is speculated might be the twenty-seven-hundred-year-old tomb of King Midas in Turkey. After archaeologists first broke into the tomb in 1957 they found the body of a man sixty to sixty-five years old lying on the remains of a wooden coffin surrounded by tables bearing bronze vessels that had contained a feast of lamb and lentils and lots of beverages to see him through eternity. They determined that the intense yellow substance in the urns was the residue of an alcoholic beverage made from beer, wine, mead, and probably saffron to add some bitterness. At a dinner in 2000, McGovern met Sam Calagione, the owner of Dogfish Head Craft Brewery in Delaware, and they discussed the possibility of making a modern version of this beverage. Calagione's creation, called Midas Touch, was introduced to the commercial market in 2001, and has since attracted a following.[8]

When McGovern asked Calagione to replicate the Jiahu grog they ran into problems finding authentic ingredients. First, they couldn't readily lay their hands on wild Chinese grapes, which grow mainly in mountainous regions, so they had to settle for a canned concentrate of the muscat grape, considered a distant cousin. They were also forced to use an orange-flower honey collected by American bees. And for rice they chose a pre-cooked gelatinized paste that had been dried, bran, husks, and all (the Jiahuans probably didn't have the technology to polish their brown rice, which may have been wild or cultivated). The next challenge was how to break down this paste into its component sugars, a process called saccharification. This is still done in many places by chewing the cooked grains and spitting them into a pot. Enzymes in saliva degrade starch into glucose, a critical part of the process of fermentation. The glucose is fed to yeast, which excretes alcohol as waste. It is also possible that the brewmasters at Jiahu added sprouted rice to the mix. Sprouted grain, called malt, also contains enzymes that reduce starch and is the most common way beers are saccharified. McGovern's group tried treating the rice paste with a traditional Chinese concoction called *qu* (pronounced "chew"), produced by using a mold to break down the starch.[9] But in the end it was the Bureau of Alcohol, Tobacco, and Firearms that dictated how the rice would be saccharified, forcing the brewers to add an ingredient, barley, the Jiahuans never used. Because Caglione intended his brew for the consumer market, he had to abide by laws that said a brewery licensed by the federal government could make beer and only beer, which had to contain 25 percent barley malt. Next, the brewers considered using only the wild yeast that is omnipresent in the environment, which is in and on everything from wild, unpasteurized honey to the skin of haws, but to speed up the fermentation process they added liquid sake yeast instead.

McGovern ordered fifty pounds of dried and powdered Chinese hawthorn online from an herbalist on the West Coast. (I'm not sure why they didn't order dried *C. pinnatifida* haws from China—available, for example, in a twelve-ounce package selling at the time of this writing for $4.29. I'm chewing on a couple of these as I write this. They're slightly sweet and acidic and have a bright, tart flavor that puckers the mouth, very different from the bland, mealy pulp of *C. douglasii* haws.) Again, the ATF intervened. It decreed that while federal law allows hawthorn to be sold as a tea or a dietary supplement, it cannot be added to beer. McGovern and his associates persevered, and the feds finally relented.

Ingredients in hand, the next challenge was figuring out their proportions. They had to rely on guesswork, since the Jiahuans did not leave behind a recipe. But after seeping the rice paste and the barley malt in water heated to 152 degrees, they were able to extract the liquid, called wort, which they fed into a kettle and boiled with the hawthorn powder. They cooled the mixture with a heat exchanger and transferred it to a fermentation tank, where it was mixed with the honey and the muscat grapes. Later in the day, they inoculated it with a liter (about 34 ounces) of sake yeast. After three weeks of fermentation the brew was ready to taste. It was too sour, and McGovern knew ancient people would never have drunk it because they loved sweet things. Adjustments were made over the course of several months, until finally they were ready to bottle their concoction. It was given a name, Chateau Jiahu. And a label, which shows an Asian beauty with bobbed hair, naked from the waist up. Turned away from the viewer, she has a tramp stamp on her lower back, a tattoo of the Chinese sign for wine, three drops of liquid emerging sideways from a symbol resembling a jar. And what did Chateau Jiahu taste like? McGovern wrote that it "hits all the right notes—an inviting, grapy nose, a Champagne-like effervescence with extremely fine bubbles, a tingling aftertaste that invited you to drink more, and a brooding yellowish color."

Whatever the original brew tasted like, it was integral to daily life at Jiahu. All of the shards McGovern tested came from residences, indicating that it was not simply a ceremonial beverage but part of the villagers' diet. Perhaps it was consumed instead of water, which might not have been safe to drink. (This was the case in the United States during the 1800s, when almost every homestead boasted an apple orchard yielding an alcoholic cider consumed by adults and children alike at meals.)[10] But alcohol no doubt played a role in the spiritual life of Jiahu, as well. To help himself achieve the sort of trancelike, otherworldly state of mind required to commune with the Unseen Force, a shaman would play his flute, bang on a drum, or shake a rattle while drinking copious amounts of grog as he escorted a dead person to the afterlife, tried a healing cure, or asked for favors from the spiritual powers. The social cohesiveness that gave Jiahu its long continuity as a community can be credited in part to the power of the shamans, but in much larger part, I believe, to alcohol. In fact, it may have been liquor that forged civilization rather than being simply a product of its dubious advance.

I say "dubious" because hunters and gatherers were generally healthier than town people. They lived longer and worked less. The food supply in a city is not as reliable as that of foragers because of crop failures due to disease or weather—consider the Potato Famine of Ireland. The cereal-based diets of city-dwellers were inferior to the game animals, fruits, nuts, and other native plants that people living in the wild ate. Infection spread rapidly in settlements where human beings were packed together. And they had to work all day to earn enough to eat, whereas foragers usually found enough food to prosper in a relatively short period of time.[11] Of course, there are exceptions, such as starvation during droughts when there's nothing available to hunt or gather. Theories abound to explain this paradox, but my favorite is that people break their backs raising grain because they use it to make alcohol.[12] Its liberating qualities cannot be denied. But since the fruit of the hawthorn played a critical role in forging the Neolithic Revolution in China, why did people farm cereals when they could have made wine more easily by throwing a bunch of haws into a vat of water and honey and going off to fish till it fermented? One reason is that there were not enough hawthorns within walking distance to make a year's supply worth of wine, especially after the birds and animals ate their fill. And either people had not learned yet how to grow more of it or their attempts failed.

Shan zha is native to the northeastern provinces around Beijing, but it is planted extensively across the mainland and Taiwan. *Shan zha* translates as "mountain berry," but to my eye the "mountain" part of the Chinese character—a horizontal line with three spikes rising from it—looks like a thorny branch. China leads the world in the harvest of hawthorn fruit, producing a million tons annually. (By way of comparison, China also produces 35 million tons of apples every year, half the world's total.)[13] The haws of the shan zha, nicknamed "red pearls," are frequently impaled on foot-long wooden skewers, dipped in boiling sugar, and sold on the street. A favorite form of this traditional treat, called *tanghulu* and dating from the Ming Dynasty, is an artful concoction resembling a bird, consisting of lines of scarlet haws topped with a transparent crest of crystalized sugar. The pulp can also be mixed with sugar and dried to produce a purplish fruit leather sold in various shapes. The fruit is used to flavor soft drinks as well. Juice from the haws produced using a cold extraction process were made into a beverage called *Qingsongling* hawthorn nectar, which became the official beverage of the Chinese delegation to the 1992 Summer Olympic Games in Barcelona. *Shan zha bing* is another Chinese treat made from the fruit.

Hawthorn is put up in a variety of teas. (I have a box of dried pulverized shan zha in tea bags that I picked up in L.A.'s Chinatown.) The seeded, sweetened pulp is formed into thin discs and dried. Called haw flakes in English, they are traditionally taken with herbal medicines that are too bitter to eat alone. (Ninety-one cases of haw flakes were returned to China in 2001 at the order of the USDA because the product contained a pink dye that was not approved for consumption in America.)[14] The most productive area in China to harvest shan zha lies in the mountainous forest region of Hebei Province, which harbors a hundred square miles of feral orchards protected from development, where some twenty other kinds of fruit, such as kiwi and pears, grow wild.

Because fresh haws tend to be sour, not all the harvested fruit is consumed, and sometimes it goes to waste. The government decreed that other uses for the resource should be developed. Since hawthorn was already being used in a wide variety of foods and herbal medicines, orchardists began looking to a relatively new sector of the economy—the wine industry. In a five-year period, from 2007 to 2012, wine consumption in China doubled twice, and domestic production has increased four times over the decade from 2003 to 2013 in an attempt to meet the demand, resulting in China's becoming the fifth-largest wine producer in the world. It is predicted that by 2016 the country will be the world's leading consumer of wine.[15] Most of this production comes from grapes native to Europe, but a niche market in hawthorn wine is also growing. While the demand for wine in general is driving some of hawthorn's popularity, Chinese consumers believe that hawthorn wine has health benefits that grape wines cannot offer, or at least not with the potency of *Crataegus.* These benefits are believed to derive from complex molecules called polyphenols, which are thought to reduce cholesterol and inflammation, retard the progress of chronic heart failure, and battle skin cancer. *C. pinnatifida* is a warehouse of polyphenols, containing some forty different compounds such as anti-oxidizing flavonoids and procyanidins, some of which interact with one another, producing synergistic effects that cannot be duplicated in synthetic drugs.[16] This is why herbalists insist that the entire plant must be consumed to get the positive effects contained in hawthorn.

A team of Chinese researchers wondered whether the process of fermentation reduces the anti-oxidant powers of hawthorn, the ability of flavonoids to snatch up free radicals. Free radicals are molecules produced by natural metabolism, by the immune system to fight microbial invaders, or by environmental damage done to cells from smoking, pollution, and the ultraviolet radiation in sunlight. A free radical is a damaged, highly unstable molecule that has lost a negatively charged electron in its outer shell and tries to stabilize itself by stealing one of these particles from another molecule in the walls of cells. When this happens the victim becomes a free radical itself. The new free radical swipes an electron from another molecule, and so on, resulting in a chain reaction that damages or even kills the cell. This kind of oxidation is the same process that an apple undergoes when it rots. The damage accumulates with age and cannot be reversed. Anti-oxidants such as flavonoids protect cells by giving up electrons to free radicals before they can attack. When this happens the flavonoid molecule turns into a free radical, but because of its chemistry it does not become reactive—that is, it does not need to replace its stolen electron to remain stable.[17]

In their experiment, the researchers fermented the fruit of *C. pinnatifida* using granulated sugar fed to five different strains of bacteria. Added sugar was necessary because this species of hawthorn has a sugar content of less than 10 percent, half as much as European grapes. Well-ripened haws with dark red skins and a fruity odor were purchased from the local fruit market in Tai'an City in Shandong Province. The seeds and pulp were separated from the juice when the yeast had eaten most of the sugar. After aging, the wine was clarified with bentonite, an absorbent clay produced from weathered volcanic ash. Then it was bottled and subjected to chemical analysis. The five strains of yeasts yielded different levels of alcohol per volume, varying from 11 percent to more than 12 percent. Researchers concluded that fermentation did not appreciably reduce the percentage of anti-oxidants contained in the fruit of

C. pinnatifida, apparently validating China's drinkers who believe that hawthorn wine is good for them.[18]

Every literate culture compiles pharmacopoeias—books listing medicinal substances and directions for their use. The Edwin Smith Surgical Papyrus, an Egyptian scroll from the Old Kingdom named for the American dealer who bought it in 1862, is one of the oldest. Others are the pharmacopoeias written by the Roman naturalist Pliny the Elder, and *De materia medica* by the Greek botanist Pedanius Dioscorides. But the first such manual produced by a government was the *Tang Bencao.* Appearing in 659 C.E., it was compiled by twenty-three experts during the reign of Emperor Xianqing of the Tang Dynasty. The fifty-three volumes of this work served as China's official pharmacopoeia for four centuries, guiding physicians through diverse medical practices and the uses of some 644 substances.

The *Bencao* was grounded in the traditional Chinese medicine that is still practiced today. In one respect it is similar to Western medicine: both traditions view disease as a struggle between hostile forces and the body. In the West it is believed that this battle is waged by microbes assaulting the immune system. The germ theory supplanted the European practice of "heroic medicine," which originated in ancient Greece and lasted until the nineteenth century. Heroic medicine was based on the belief that maladies were caused by an unbalanced state of the "four humors"—blood, yellow bile, black bile, and phlegm. The agent that triggered this unbalance was called a miasm, a cloud of polluted air generated by putrefied organic matter. Balance was restored by barbaric treatments that did more harm than good. These included purging, bloodletting, sweating, and blistering. Purging involved the administration of mercury chloride, or calomel, to make the patient vomit. Bloodletting probably killed George Washington: half his blood was drained in a single day to treat an inflamed epiglottis. Because this widespread practice lost favor among doctors, barbers, who also functioned as surgeons, became its chief practitioners, hence the barber pole—red for blood, white for bandages, and the pole to symbolize the rods that were squeezed to force the blood from the arms. Science gradually replaced superstition as clinical trials and serendipity revealed new drugs and methods. What is called evidence-based medicine—treatments based on what has been shown to heal—replaced trial and error and educated guesses.[19]

According to traditional Chinese medicine, disease is caused by a disharmony within the body or between the body and the environment resulting from the struggle between antagonistic forces and human energy. At the core of the tradition is the belief in yin and yang, seemingly opposite phenomena that are bound up with one another: the top of a hawthorn tree grows in the light, for example, while its roots grow in the dark. The forces that act on the human body include six atmospheric or climatic forces such as cold and dryness; seven emotions; and the forces of fatigue and food. Energy, or *chi,* flows inside the body along what are called meridians. These energy highways are thought to be channels that carry, hold, and transport chi, blood, and bodily fluids. Traditional Chinese medicine has been dismissed by many scientists in the West, but its concept of disharmony as the cause of disease goes farther than germ theory, for example, in explaining why the virus responsible for chicken pox, which lives in the nerve endings of most adult Westerners, can become reactivated and cause shingles.[20]

In the *Bencao,* shan zha is considered useful for treating "food stagnation," that is, sickness brought on by overeating. The treatments contain hawthorn mixed with various other components such as forsythia (an Asian herb that was touted as the cure for the viral plague portrayed in the movie *Contagion*). One drug is called the Great Tranquility Pill. Another is the Major Cratageous Pill. As an herbal treatment for digestive problems hawthorn is often combined with a grass called *Hordeum* and a concoction of herbs and grains called *Massa fermentata.* Taken together, these are known as the "Three Immortals." The manufacturers of the Brain-Vitalizing Pill, which is mostly *Crataegus* with a few other ingredients such as the genus *Carthamus,* which contains the species safflower, claim that their product will make users smarter.[21] The *Bencao* also credits hawthorn with relieving diarrhea. It is believed that the essence of fruit from *C. pinnatifida* enters the stomach and spleen "channels," and the organs themselves, stimulating the blood to improve digestion, promote appetite, and remove accumulated food.

Shan zha is also given in pill or powder form to animals. A favorite story about hawthorn is set during the Ming Dynasty. On the advice of an incompetent village doctor, a mother began overfeeding meat to her jaundiced boy. He had trouble digesting the big, rich meals and his abdomen began to swell. Seeking relief from the pain, he wandered into the mountains, where he found a tree bearing small red fruits. Eating one, he discovered how delicious they were and ate more. After he got home, he vomited phlegm. His belly shrank, he gained weight, and the bloom of good health returned to his face. (This is how *C. pinnatifida* got its name of mountain berry.)

Intertwined with traditional Chinese medicine, Daoism is a spiritual and ethical philosophy embracing the belief that massaging the organs leads to better health and an improvement in chi. The theory, *chi nei tsang,* embraces a school of thought which holds that communing with certain trees concentrates the masseuse's powers. Trees not only animate the world by converting carbon dioxide into oxygen, they can also transform bad energy into good. Believers therefore go out into city parks and hug them. Although all trees are in a constant state of meditation, members of the urban forest are believed to be less skittish and more willing to share their energy than wild trees in the countryside. Different kinds of trees offer different therapeutic benefits. Communing with a willow, for example, will help the therapist expel the patient's sick winds, rid the body of excess "dampness," and strengthen the bladder and urinary tract. The hawthorn aids digestion, strengthens the intestines, and lowers blood pressure.

After he came to power in 1949, Mao Zedong championed a resurgence of traditional Chinese medicine and the village healers, called "barefoot doctors," who administered it, although he chose to ignore Daoism's ancient role in health care, a convenient contradiction, because he found it impossible to divorce the spiritual from the medical. The Chairman was motivated by economics: any revolution that promised to lift up masses of peasants would have to provide health care, but the Western, evidence-based medicine practiced by a few doctors in China at the time was something the entire country could not afford. At Mao's direction herbalism and acupuncture were revived. (*Under the Hawthorn Tree* includes a series of events centering on a healer's prescription of walnuts and rock sugar to treat a problem Jing Qiu's mother had with blood in her urine.) In public he declared that the power of

Chinese medicine was responsible for the country's robust population growth. In private, however, Mao admitted that he did not believe in Chinese medicine and would never allow it to be practiced on him.[22]

Little was known in the West about Chinese medicine until the journalist James Reston wrote in the *New York Times* in 1971 about the appendectomy he underwent at the Anti-Imperialist Hospital in Beijing while visiting the country with his wife. Following the surgery he was in considerable intestinal pain, probably due to gas. An acupuncturist was called in. He inserted long, thin needles in Reston's elbow and behind his knee on the meridians that were thought to govern the bowel. Within an hour, Reston wrote, the pain and bloating in his abdomen eased and never returned.[23] This ignited a mania in the West. People clamored to get into acupuncture seminars. The U.S. government reversed a former ban and ruled that large quantities of needles could be imported from China. The acupuncture craze was further fueled by a case involving open-heart surgery in China using acupuncture as the only anesthesia. Enthusiasm for the practice waned when this was later proven to be a hoax—the patient had actually been given powerful synthetic pain killers. Other sham operations were exposed. Within a decade the craze in America was over.

The twelve meridians have a rough equivalent in the nervous and circulatory systems, but there is no clinical evidence that these energy pathways exist. They were invented by the ancient Chinese, who had little direct knowledge of anatomy because they were forbidden by Confucianism to dissect cadavers, which constituted a defilement of the human body.[24] Some doctors and scientists in the United States thought that acupuncture might have validity, however. Rigorous clinical trials were designed and carried out in American universities to study the effects of needles on such non-life-threatening conditions as lower back pain, angina, migraines, and arthritis. In 1979 the World Health Organization announced after studying these trials that more than twenty maladies ranging from tennis elbow to dysentery had responded well to acupuncture. The WHO revisited the question of acupuncture in a 2003 report and concluded that there were twenty-eight conditions, ranging from hypertension and morning sickness to depression and adverse reactions to chemotherapy, for which acupuncture has "been proved—through controlled trials—to be an effective treatment." Critics argue that because of two significant errors the report is deeply flawed. First, it took into account *too many* clinical trials, skewing the results by including poorly designed studies along with good ones. Second, the results of clinical trials in China should not have been included. Because the Chinese pride themselves on their traditional medical practices, critics allege that researchers succumbed to publication bias—the eagerness to report only trials with positive outcomes. Their findings were simply too good to be true, the naysayers said, and were not verified by similar trials in the West.[25]

As researchers in China were contributing these allegedly flawed results of clinical trials testing acupuncture, they also began publishing studies investigating the efficacy of traditional botanicals to treat medical conditions. Since hawthorn plays a small but notable role in the culture and economy of China, it is one of the plants that has been subjected to such tests. In a controlled clinical trial with ninety-two patients suffering from angina, chest pain caused by heart muscle that is not getting enough oxygen, for example, it was reported that *C. pinnatifida* extract decreased the number of attacks by 85 percent compared to 37 percent for the half of the sample getting a placebo, and it improved the electrical functioning of the heart as measured by an electrocardiogram by 37 percent in patients getting hawthorn as opposed to 3 percent for the placebo group.[26] A 2009 study concluded that hawthorn extract lowers the level of artery-clogging fat in the bloodstream. Other Chinese researchers have analyzed the chemical components of shan zha and dissected rats to see what effect flavonoids extracted from hawthorn had on the tissues.

So as the American medical community reluctantly embraced acupuncture for some maladies, the Chinese were adopting the research protocols of evidence-based science. However, while it seems that Western and traditional Chinese medicine are courting, no date has been set for the wedding. But the epidemic of heart disease has compelled clinical researchers around the world to look at ancient medicines, such as hawthorn compounds, that might treat these deadly ailments without killing the patients.

EIGHT

The Medicine Tree

> These are the drugs that, had they been available, some presume might have saved many of my father's generation. But far from saving either their lives or ours, clinical trials show that the indiscriminate use of drugs to lower lipids or blood sugars, to relieve respiratory wheeze, or to block stress hormones may even increase the risk of loss of life, and appear to be doing so in the United States.
>
> —DAVID HEALEY

The odds are that you and I will die of heart disease. In 2011 some form of the malady killed 17 million people across the globe, and heart disease was the leading cause of death worldwide. During that year cardiovascular ailments claimed 750,000 lives in the United States, cost $273 billion in medical care, and deprived the economy of $172 billion because of premature deaths and days lost at work. In Europe four million people die every year of these diseases, which cost the European Union $269 billion annually. Despite the fact that the rates of heart disease have been declining for thirty years, mostly due to the souring of our love affair with the cigarette, these figures are projected to rise significantly by 2030. In that year more than 40 percent of the aging American population will probably develop some form of heart disease, including high blood pressure, heart failure, coronary heart disease, and stroke.[1] Although the essence of hawthorn has been used to treat cardiovascular problems in the West for at least five hundred years, the medical community in the West only began in the early 1980s to rigorously investigate whether there is any scientific basis for the old folk belief that *Crataegus* is the tree of life.

The first mention in the West of hawthorn used as a remedy appears in the five-volume *De medica materia,* compiled by Dioscorides. "Taken in a drink or eaten," he wrote in first-century Rome during the rule of Emperor Nero, "the fruit stops stomach outflows and the flows of women. The root bruised small and applied draws out splinters and thorns. It is said that the root is able to cause abortions, the stomach being touched gently with it or rubbed with it." He also advised parents who wanted to bring a male child into the world to drink a concoction of hawthorn seeds in water for forty days before having intercourse. Dioscorides not only wrote about some six hundred plants, which he came across during his extensive travels serving as a surgeon in Rome's imperial army, he described the medicinal uses for some seventy-five creatures, as well. The testicles of beavers, he advised, are "good against the poisons of snakes." Small pieces of bedbug inserted into the urethra would cure painful urination. Despite much ludicrous advice and hand-me-down myths, his work has always been known, unlike many other classic works from antiquity that were lost for centuries and then rediscovered during the Renaissance. In fact, *De medica materia* constituted the bulk of Europe and America's pharmacopoeia well into the 1800s.[2]

Joseph Du Chesne, a French alchemist and one of Henri IV's physicians, wrote in 1603 that a syrup made from the fruit of the hawthorn was effective in treating heart problems. (It is unknown whether His Royal Highness needed a cardio tonic; what he certainly needed was a bath. It was reported that he smelled so bad his wife-to-be, Marie de' Medici, fainted when first they met.)[3] Du Chesne was influenced by a Swiss physician named Paracelsus, who believed that instead of blindly following the teachings of ancient texts and physicians such as Dioscorides, a doctor should seek the truth by studying the natural world and the effect of his treatment on the patient. At the time this was considered radical thinking. A relentless self-promoter, Paracelsus allegedly drew attention to himself by publicly burning the *Canon of Medicine,* the standard textbook used in medieval universities. At the same time, he believed in the ancient Doctrine of Signatures, in which a doctor could ascertain what a plant was good for just by looking at it: herbs effective against jaundice were yellow flowers such as marigolds and dandelions; red hawthorn berries suggested that they could be useful in the treatment of blood and heart ailments. And the best way to remove thorns or splinters from a puncture wound was to soak a cloth in a decoction of thorny twigs and apply it to the wound. As the English botanist Nicholas Culpepper claimed in his 1653 *Complete Herbal,* "the thorn gives a medicine for its own prickling."

Du Chesne was probably guided to the hawthorn by the collective experiences of amateur healers. Usually women, these village herbalists have been treating people for millennia. In the fall, for example, Native American women in the west gather roots from a spindly mountain plant two to three feet tall. *Ligusticum porteri,* commonly known as "bear root" or "bear medicine," is used as an anti-bacterial, anti-viral agent. Mixed with honey it is taken to treat coughs and bronchial maladies. It is believed that Indians learned about the plant from watching sick bears, which eat almost everything, as they tried to cure themselves by digging up the plant and eating its roots. Both bears and people have always had to be careful, however, because the plant looks very much like *Conium maculatum,* poison hemlock, which paralyzes the central nervous system.[4]

How hawthorn made its way into the healer's basket is not known. While its chemistry is complex, no active ingredients have been isolated. Even if hawthorn is proven to be as efficacious as some traditional pharmacopoeias have claimed, doses that an animal or a human might get by nibbling the leaves, flowers, and berries are low and would work only after several months of daily use. These days highly concentrated extracts are prescribed for heart patients in Europe. Unlike the relatively fast action of plants such as marijuana, for example, or cranberries in treating bladder infections, hawthorn is so slow-acting that healers and sufferers are unlikely to notice any difference for some time. It is more likely that because the tree is edible and mildly tasty people in antiquity ate from it throughout the growing season and may have eventually discovered that their hearts had become stronger and they felt more robust.

In the late 1800s an unlicensed Irish doctor named Greene in Ennis, County Clare, was said to be successfully treating his heart patients with a secret formula that gained him wide notoriety. Although professional ethics compelled him to share the ingredients, he chose not to because he was making money from it. After his death in 1894, however, his sole heir, a daughter known to us only as "Mrs. Graham," revealed that this elixir was a tincture compounded from the ripe fruit of "*Crataegus mon-*

ogyna," which at the time referred to several other hawthorn species in addition to *C. monogyna*. Well, maybe. It's a good story, embraced and retold by American homeopaths and herbalists for more than a century, and it might even be true, although I couldn't find a "Dr. Greene" among the headstones in County Clare's twenty-three cemeteries. In 1896 an American physician named M. C. Jennings published a strange, unsubstantiated letter in the *New York Medical Journal* about "Dr. Greene" and his miraculous cure. Jennings claimed that he had been using the Irishman's formula to treat patients of his own, and included case studies to illustrate its efficacy.

Among the 118 heart patients Jennings claimed to have treated successfully was a young woman whose family summoned him after finding what appeared to be her lifeless body. "I went in and found that she was not quite dead. . . . I gave her hypodermically ten drops [of the hawthorn elixir], and in less than half an hour she was able to talk and describe her feelings." He also detailed the case of a woman from Louisville who was suffering from an enlarged heart. A "faith-cure man" had been administering to her, Jennings reported, but under his care her condition worsened. Jennings treated her with Greene's formula, along with digitalis, a derivative of foxglove that is a proven agent and is still widely used to control heart rate and atrial fibrillation. The woman soon recovered enough strength to go home and wrote Jennings three months later that she was feeling fine. He also claimed to have cured a man of seventy-three who was gasping for breath and had a pulse rate of 158. After Dr. Jennings dosed the man with fifteen drops of *Crataegus* his pulse rate dropped and his breathing became easier. "He made a rapid and apparently full recovery," Jennings wrote, and after continuing to take hawthorn, "in three months, he felt as well as any man of his age in Chicago." Hawthorn was superior to any of the other treatments for heart problems, Jennings concluded, "because it seems to cure while the other remedies are only palliative at best."[5]

Whether Jennings' case studies were fact or hyperbole, tincture of *Crataegus* soon became one of the most widely used cardiovascular treatments in America. In the late 1800s stories about hawthorn's triumphs began showing up in medical journals. A Kansas City doctor name Joseph Clements reported a case in 1898 of a man who had been suffering excruciating attacks of angina. His heart felt as if it were "gripped by bands of iron," producing an "overwhelming sense of coming calamity or dissolution." The man took six to ten drops of tincture of *Crataegus* four times a day for several months. The result, Clements reported, was that every symptom of the disease disappeared.[6]

Among those who were eager to try the remedy were physicians from the American School of Medicine (Eclectic), which trained doctors in a dozen privately funded facilities largely in the Midwest. These physicians became known as the Eclectics. Their theories and treatment methods arose during the 1840s in reaction to the "heroic medicine"—including bloodletting and purges—used by most doctors at the time. Seeking to treat the person and not thc pathology, which is to say that what works for one individual might not work for another, Eclecticism was the extension of the ancient traditions of herbalism in Europe and America. It borrowed from the extensive pharmacopoeias of Native Americans and incorporated the experiences of midwives, who for centuries had been gathering herbs and concocting treatments in addition to delivering babies. Largely basing their practices in rural communities, the Eclectics believed that one of the keys to a successful remedy was the support of a patient's "vital force." Although neither chemical nor mechanical, this unseen effervescence, they claimed, governed human health.[7]

Renowned among the Eclectics was Dr. Finley Ellingwood, who published numerous professional endorsements of *Crataegus* in his *American Materia Medica,* including the pronouncement from a Dr. Jernigan that the tonic "dispels gloomy foreboding, increases the strength, regulates the action of the heart, and causes a general sense of well-being." Other physicians claimed that hawthorn was also useful in the treatment of goiter, asthma, and kidney disease. "There can be no question as to its value as a tonic to the heart-muscle," wrote another noted Eclectic in 1922, Dr. Harvey Wickes Felter. However, Felter was reluctant to endorse what he considered a new treatment: It was, he noted, "still on trial; and as yet with no rational explanation of its reputed powers."[8]

American pharmacists who compounded medicines for physicians probably bought their *Crataegus* preparations from the firm of Lloyd Brothers in Cincinnati. By 1884 the company was manufacturing 835 tinctures, called "fluid extracts." But the core of its business was a line of "Specific Medicines," which were geared toward Eclectic physicians. These concoctions were highly concentrated tinctures eight times the strength of those usually prescribed by the medical community. Most were produced by first macerating, or softening, a botanical in a solvent such as alcohol, ether, or water, then extracting its essence. Instead of subjecting all medicinal plants to the same process it was believed that each one had unique properties that required unique methods of extraction. In the case of *Crataegus,* the fruit was soaked in neutral grain alcohol that had been distilled several times to remove impurities. The result was a brown-red liquid said to have a fruity, wine-like aroma and a pleasant, slightly acidic taste. According to sales material the company released, only the fruit had potency: the bark and the root were "therapeutically valueless." While Lloyd's "fluid extracts" were prepared from the haws of *C. monogyna* imported from England, when the company began preparing a Specific Medicine from hawthorn, the publicity materials explained, "comparative investigations finally led us to the conclusion that this berry is inferior to one of the American species." The identity of this species was never revealed, however. Lloyd Brothers claimed that its hawthorn elixir was "a curative remedy for organic and functional heart disorders, including cardiac hypertrophy, with mitral regurgitation from valvular insufficiency, and angina pectoris. Sometimes spinal hyperemia is associated with the latter, when both are said to be relieved by the drug." Best of all, the company promised that the hawthorn extract could be taken in large doses for a long time without producing any ill effects.[9]

Extractions were carried out in a "cold still" that exposed plant matter to a minimum level of heat, which tends to degrade the extracted material. The still was also designed to preserve as much of the solvent as possible in order to reduce manufacturing costs. This singular device, patented in 1904 and still in limited use today, was so effective that Johnson and Johnson used it to process 150,000 pounds of belladonna annually in the early years of the twentieth century. (Also called deadly nightshade, *Atropa belladonna* is the source of the drug atropine, which is critical in so many remedies in addition to its use to treat heart conditions that the World Health Organization has declared it a core drug in its list of essential medicines for every health care system.) Lloyd's still was invented by John Uri Lloyd, the oldest of the three brothers who founded the company in 1886. When he was fourteen Lloyd was apprenticed to a chemist, at the time the prevailing means of getting an education in the sciences. His brothers became chemists, as well. When he was twenty-one he was hired by two Eclectic doctors to evaluate their pharmaceutical preparations. Later, Lloyd would teach at the Eclectic Medical Institute and the Cincinnati College of Pharmacy, where he prohibited his students from taking notes, requiring them to memorize his lectures.[10]

John Uri Lloyd died in 1936 at the age of eighty-one just as Eclecticism was being forced from the scene. Part of its fall from favor came at the hands of its rival, the American Medical Association, which asked the Carnegie Foundation to commission a study of U.S. medical schools to determine whether they were meeting AMA standards for educating doctors. The foundation gave the job to Abraham Flexner, who was neither a doctor nor a scientist but was rather the owner of an experimental for-profit high school in Kentucky. Released in 1910, the Flexner Report declared that there were too many medical schools graduating too many poorly trained doctors. To achieve a reduction in numbers and an elevation in quality the report recommended increasing the prerequisites required for admission and graduation, and closing private medical schools or merging them with universities. Because homeopathic schools were "a striking demonstration of the incompatibility of science and dogma," Flexner wrote, a uniform curriculum should be established that was based strictly on mainstream science, or what the Eclectics sneeringly called allopathy—the use of physical or pharmacological intervention to suppress the symptoms of disease. As a result of the Flexner Report, the number of medical schools fell from 150 to 81 in 1922. (Currently there are 171 accredited schools in the United States.) The number of doctors graduating dropped in the immediate years following the report from 4,500 to 3,500. But another reason Eclectic doctors lost ground was the triumph of the germ theory of disease and the rise of wonder drugs such as penicillin and insulin. (Some critics believe that the Carnegie Foundation and the AMA ushered in government control of American health care in order to restrict access to the increasingly lucrative field of medicine. The physician, libertarian, and former Texas congressman Ron Paul charged that the report's real motivation was to shut down medical schools that catered to women, minorities, and homeopathy.)[11]

By the 1920s the Eclectic Medical College, as the institute was renamed in 1910, was attracting fewer students and facing increasing problems with accreditation. In fact, there were no graduations from 1929 to 1933. In 1935 a medical examiner who toured the college concluded that it did not have adequate equipment, suitable facilities, or enough instructors, and recommended that the school's application for accreditation be denied. Since the college had lost its prime source of financial backing at the death of John Uri Lloyd, it closed its doors in 1939, the last survivor of two dozen schools of Eclecticism in America.

Lloyd Brothers, meanwhile, had been sold to the S. B. Penick Company in 1938, but the trustees of John Uri Lloyd's estate, who included his son John Thomas Lloyd, apparently failed to turn over the Specific Medicine formulas to the new firm. When John Thomas founded his own company, John T. Lloyd Laboratories, and began manufacturing a line of pharmaceuticals called Lloydson Medicines, S. B. Penick sued, alleging that he had stolen the books recording the formulas. But the only thing that came of the suit was a note Lloyd was compelled to place in his advertising stating that his company had no connection to his father's business. While no one has proven who took the formula books, many believe that John Thomas Lloyd was indeed the culprit. Their whereabouts remain a mystery to this day, the Holy Grail of herbalism, inspiring myriad speculations about botanical miracles that lie waiting to be discovered among John Uri Lloyd's formulas.

America's interest in the hawthorn faded. But in Britain and Ireland, where the tree has been intimately linked with the political, cultural, and religious history of the islands for centuries, its use as a medicinal plant continued. In 1939 a researcher at the University of Glasgow reported that he had conducted a series of laboratory tests featuring a tincture of *Crataegus* administered to anaesthetized cats, rats, guinea pigs, sheep, dogs, and rabbits. He found that hawthorn slowed the heartbeat of the mammals, but constricted their coronary arteries. Treating several human subjects suffering from heart problems, he concluded that "there was no improvement in cardiac function." Next, he treated ten people with high blood pressure. "In every case the systolic and diastolic pressure was reduced, often sharply," he reported, "and returned to its former level some fourteen days after stopping medication." Although this was an early attempt to quantify the effects of *Crataegus* on the cardiovascular system, the experiment and its conclusions have little validity because the sample of patients was too small, it was not a blind experiment, and the sample was not randomized.[12]

In clinical trials used to evaluate the usefulness of drugs to treat people it is critical to prevent the introduction of bias on the part of the patients or the researchers. Otherwise, the results will be compromised. Since this bias could be unconscious all information about the experiment is withheld from the participants. In a double-blind trial both those tested and those doing the testing are kept in the dark. In a single-blind experiment the testers are privy to information; single-blind tests are sometimes unavoidable because of the nature of some experiments. (One example of a single-blind test that some critics argue should be double-blind is the police lineup. Because the police know who the suspect is, and their prejudice might inadvertently influence the witness, those administering the lineup ought to have no knowledge of the suspect.)

One of the first single-blind trials on record was carried out in 1784 to assess the claims of a German physician named Franz Mesmer that he had discovered something he called "animal magnetism." Using magnets, penetrating stares, pressure on the hands and arms, and evocative music, Mesmer believed that he could shift a mysterious fluid or "tide" inside the human body and bring about healing. Called Mesmerism, his theory held that illness occurs if the flow of these tides is blocked. When the tides were freed the patients suffered a momentary "crisis" before the catharsis of relief. A man being treated for madness, for example, might suffer a brief psychotic episode before his mental health was restored. But sometimes Mesmer found that his subjects fell into a trance.[13]

Mesmerism had acquired a huge following in Europe, attracting people from all classes—peasants, wealthy hypochondriacs, and such luminaries as Mozart and Marie Antoinette. Rumors that Marie Antoinette was engaging in sexual orgies at Mesmer's residence compelled

Louis XVI to order an investigation of the theory. The experiment took place in Paris at the residence of Benjamin Franklin, who was then America's minister plenipotentiary to France. Franklin was appointed chairman of the commission of scientists observing the trial because of his experience with electrical current. Mesmer did not attend but sent an associate, Dr. Charles-Nicolas d'Eslon, to conduct the experiments. D'Eslon told some subjects they were being magnetized when in fact they were not; he told others that they were not being magnetized when they were. Those who believed they had been magnetized, even when no magnet was used, reported feeling better. The subjects who had been secretly magnetized showed no improvement. (This was one of the first recorded uses of a placebo.)

Because he was in such demand, Mesmer had "magnetized" trees growing in public places and tied his patients to them with ropes, being careful to avoid knots that might block the flow of the tides. Although *Crataegus* is a common tree in the urban forest of Paris, and it was the custom for French mothers to take their ailing children to a flowering hawthorn and beg the tree to restore their health, any record of the species of trees that were mesmerized has been lost. To test Mesmer's idea, d'Eslon was instructed to magnetize an apricot tree in the garden of Franklin's residence. He passed by the first four in a row and magnetized the fifth. A sickly, blindfolded boy of twelve was led to each of the trees, which grew a considerable distance apart. At the fourth tree the boy fainted. The commission concluded, "The magnetic fluid does not exist, & the means used to activate it are dangerous."[14] It was also deemed that the positive outcomes were the result of "the imagination." Mesmer left France the next year, and eventually faded from the public eye, leaving behind his legacy: the discovery of hypnotism.

In order to assess the usefulness of a drug, its cost-effectiveness, and any adverse side effects, researchers rely on randomized controlled trials (RCTs, also called randomized clinical trials). In medicine, a double-blind RCT is considered the gold standard of scientific experimentation. To use a simple example, to determine the consequences of administering an extract of *Crataegus* to people with a heart condition, experimenters would first round up a suitably large population of sufferers with roughly the same diagnosis. Several hundred would be acceptable; several thousand would be better. The extract would be administered to Group A, a placebo to Group B, and digitalis to Group C. Each substance would be put up in an opaque yellow gelatin capsule to mask its contents. In order to prevent intentional or subconscious bias in the selection process, the people would be placed at random into the three groups, perhaps through assigned numbers that a computer program such as Excel would generate. To further ensure that no bias tainted the trial neither the subjects nor the clinicians administering the experiment would know which capsules contained which substance.

Another kind of bias some researchers believe can creep into a clinical trial results when one group in the sample becomes aware of being monitored more rigorously than the others. The phenomenon, called the Hawthorne Effect, was named after the Hawthorne Works, an enormous manufacturing complex near Chicago owned by the Western Electric Company, which produced most of America's telephones during the first decades of the twentieth century. It was located in a subdivision that real estate developers in the nineteenth century named Hawthorne, a common spelling at the time for the trees growing there.[15] These were probably Washington hawthorns, *C. phaenopyrum,* which are native to Illinois. (Developers like to name their subdivisions after trees to make them sound more attractive, but they often cut down the namesakes to clear space for more houses.) In 1924 researchers arrived at the plant to determine whether better lighting increased the productivity of the workers. After increasing the lighting, they found that productivity also increased. But when they reduced the lighting to its original level they were surprised to discover that productivity increased as well. Over the next nine years they continued to tamper with other workplace conditions and claimed to have gotten the same result: productivity always increased. According to psychologists analyzing the data five decades later, "Regardless of the conditions, whether there were more or fewer rest periods, [a] longer or shorter workday . . . the women worked harder and more efficiently." Their explanation was that the alterations in workplace conditions were not responsible for the increased productivity; the attention paid to workers by the people who were monitoring their performance was. They were proud that they were be part of an experiment, the social scientists concluded, so they worked harder. In fact, the social scientists were also mistaken. The original experiment failed because its methodology was deeply flawed: it had only five workers as subjects, and two were replaced midway through. Other workers increased their productivity as they became more accomplished because they were getting paid by the piece, not by the hour. Since it was first conducted, no one has been able to duplicate the results, further proof that it was a poor experiment.[16]

Results of the first recorded randomized experiment were published in 1885 by Charles Sanders Peirce, an American logician whose theory that certain operations could be performed by gates in an electric circuit eventually led to the invention of the computer. In his experiment, titled for publication "On Small Differences in Sensation," a researcher sat on one side of a screen and a subject on the other. Inserted into an opening in the screen was a post office scale adapted for the experiment. As very small weights were added to the scale it put pressure on the finger of the subject. As they were removed the pressure decreased. The subject was asked to describe what he was feeling. Peirce determined whether to add or subtract weights by selecting a playing card from a shuffled deck—thus the randomization. He believed that the results showed that there is no amount of pressure, regardless of how little, that his subjects could not feel. He concluded that his experiment validated the existence of telepathy: "It gives new reason for believing that we gather what is passing in one another's minds in large measure from sensations so faint that we are not fairly aware of having them." This ability also accounted, he believed, for "the insight of females."[17]

The first randomized clinical trial was carried out in 1946 in England to investigate whether streptomycin, which was then a new antibiotic discovered by a graduate student at Rutgers University in 1943, could be used to fight pulmonary tuberculosis. In the trial, 107 patients with an untreatable form of the lung disease were selected. Bed rest was prescribed for 52 of these patients while 55 were treated with streptomycin injections and then prescribed bed rest. (The number of subjects was restricted because at the time the antibiotic was in limited supply.) Researchers decided that to control the experiment they ought to minimize the variables. The patients selected ranged in age from fifteen to thirty, the onset of their tuberculosis was recent, and it was unlikely that they could defeat the disease on their own. Each patient was assigned a number from a set of random numbers drawn up for each sex. The trial was conducted double-blind: the details of each case were kept in sealed envelopes labeled with the patient's number, and the patients were not told that they were part of a study. Nor were the doctors told that they were taking part in a trial.

The patients remained in bed for six months, then X-rays of their lungs were examined by radiologists who also knew none of the details of the trial. Before the end of six months four of the patients receiving streptomycin had died, as did fourteen of the patients in the control group (the group that did not receive the drug). The condition of those treated with the antibiotic improved more than that of patients who only got bed rest. Eight of those taking the antibiotic were cured. At the same time, in the latter part of the trial the bacteria developed a resistance to the antibiotic, causing a worsening in the condition of the drug-takers. [18]

Although hawthorn has long been used as a cardio tonic, it was not until the 1980s that scientists began testing its use for treating heart failure. In 2001 a German researcher published the results of a randomized, double-blind clinical trial designed to see whether *Crataegus* had any effect on people with mild chronic heart failure. In the study twenty patients were given a 240-milligram capsule of a *Crataegus* extract three times a day, and twenty were given a placebo. Both sexes were included in the trial. After twelve weeks patients were asked to work out on stationary bikes. The group that took the *Crataegus* showed improvement in what is called "exercise tolerance," while the tolerance of the placebo group declined. The heart rate of the hawthorn group also lessened, as did their systolic blood pressure (the higher of the two numbers in a measurement, indicating the pressure when the heart has just pumped). No adverse reactions to the extract were observed. While the results seemed promising, because the sample was so small other researchers questioned whether the trial had any significance.[19]

Congestive heart failure is a serious condition in which the cardiac muscle, called the myocardium, has weakened and can no longer keep up with the body's demand for oxygen. The heart tries to compensate, by getting bigger, which allows it to stretch more and contract more strongly so it can pump more blood. It might also develop more muscle mass because the cells that govern its contractions grow more robust. Or it might begin pumping faster to increase its output. Vessels might narrow to compensate for the organ's diminishing power, or the body might divert blood from less important tissues in order to make sure the brain and the heart get as much as they need. These strategies work for awhile, possibly for years, masking the condition. But finally the heart can no longer do its job.

There are four stages of chronic heart failure. As defined by the New York Heart Association and accepted internationally, they range from Class I, in which the symptoms, such as fatigue and dyspnea (labored breathing), have not yet appeared, to Class IV, in which the symptoms appear even when the patient is at rest, and physical activity leaves the patient breathless, anxious, and uncomfortable. A virtual medicine cabinet of drugs is prescribed for the condition, which is incurable. These include blood thinners, diuretics, digitalis, beta blockers (to lower blood pressure or treat chest pain), and calcium antagonists (to relax blood vessels and correct abnormal heart rhythms). The unpleasant side effects the various drugs can cause sound like the litany of warnings you hear in television commercials that make you think you'd rather be dead than use any of the drugs. According to the American Heart Association, these include but are not limited to dizziness, headache, constipation, diarrhea, nausea, indigestion, depression, insomnia, hot flashes, and loss of sex drive.

My father died of congestive heart failure when he was seventy-five. Although none of his doctors treated him with hawthorn because its therapeutic value has only recently come under consideration in America, in Europe he would probably have been put on a *Crataegus* regimen in addition to a synthetic drug or drugs. In Germany these extracts are prescribed specifically to treat Class I and Class II cardiac failure. They were approved by the Kafkaesque-sounding Kommission E, a panel of scientists that advised the German equivalent of the U.S. Food and Drug Administration about substances that had traditionally been used in herbal and folk remedies. Between 1984 and 1994 the Kommission published a series of monographs that summarized the safety and usefulness of 308 herbs that could be prescribed legally by German doctors. The

Crataegus monograph, which was released in 1994, contains a report that extracts of hawthorn leaves and flowers, macerated with alcohol and water, are useful in treating Stage II cardiac failure. The Kommission concluded that the herb promotes an increase in the patient's ability to exercise, a decrease in blood pressure and heart rate, and an increase in the amount of blood pumped from the heart with each beat. In addition, it generally has no side effects. These finding were based on experiments performed in test tubes or on animals.[20]

A study with much more statistical significance than the 2001 study was also designed by German researchers. Conducted in 2008, it was called the SPICE trial (Survival and Prognosis: Investigation of Crataegus Extract WS 1442 in congestive heart failure). Almost twenty-seven hundred patients with Class II and Class III cardiac failure were randomly divided into two groups: one group received nine hundred milligrams of hawthorn per day for twenty-four months, the other received a placebo. In general, the results indicated that hawthorn was not effective in preventing deaths, hospitalizations, or nonfatal heart attacks, but they suggested that *Crataegus* might reduce sudden deaths in patients whose hearts are pumping blood at a rate of 25 to 35 percent of normal.[21]

In 2002 a randomized sample of 209 patients with Class III heart failure was divided into three groups, those receiving an eighteen-hundred-milligram capsule of hawthorn extract every day, those getting nine hundred milligrams, and those getting a placebo. At the end of sixteen weeks they worked out on stationary bicycles. The results demonstrated that for patients receiving the higher dose their capacity for exercise had improved and their symptoms had been reduced.[22]

In 2004 a modest study was conducted in Iran involving ninety-two patients aged forty to sixty with high blood pressure. The subjects were randomly divided into two groups, one receiving hawthorn and the other a placebo three times a day. After four months both the systolic and diastolic blood pressure readings of the hawthorn recipients had significantly lowered. In a 2004 trial conducted at the University of Michigan, 120 patients were randomly divided into two groups. One received nine hundred milligrams of hawthorn once a day and the other received a placebo. All the patients were also being treated with the usual arsenal of synthetic drugs. After six months the results showed that more people who received hawthorn had died or required hospitalization than those taking the placebo. Despite the negative and inconclusive outcomes, in a 2010 study of the trials summarized above, as well as six other studies, scientists in Ireland concluded that overall the results indicated that "*Crataegus* preparations hold significant potential as a useful remedy in the treatment of CVD [cardiovascular disease]." Even when the hawthorn did not help, it did not cause any serious side effects and it did not react negatively with any other drug.[23]

The results of the Michigan study, the report concludes, "have no rational explanation and have not been observed by other studies." One of the Michigan researchers conceded that the higher rate of deaths from heart failure in the group receiving hawthorn might have been due to chance, citing the small size of the sample. And the results might have been skewed by the fact that the patients were in the second to fourth stages of cardiac failure and taking a plethora of other drugs. Nonetheless, Dr. Keith Aaronson, the principal investigator of the Michigan trial, told me that he was no longer interested in studying the therapeutic possibilities of hawthorn. Because of budget restrictions on the government agency that sponsored the study, the National Center for Complementary and Alternative Medicine, it is possible there won't be any more U.S. clinical trials of *Crataegus*. The agency's 2013 budget amounted to significantly less than 1 percent of the budget of its parent agency, the National Institutes of Health.[24]

Which of the ten thousand biologically significant phytochemicals occurring in the world's plants might account for hawthorn's reputed therapeutic benefits? The extract most often used in clinical trials of *Crataegus* is a commercial product called WS 1442, manufactured by a German firm, Dr. Willmar Schwabe Pharmaceuticals. According to the company, it harvests the leaves and flowers of *C. monogyna* from its own plantations and processes them into a dry extract in capsule form. This has been "standardized" to contain 18.75 percent of a flavonoid called oligomeric procyanidin. Standardization of plant extracts guarantees that the phytochemicals thought to provide remedies remain the same for each batch.

Flavonoids, recall, have been shown to neutralize free radicals, the unstable molecules in the body that can damage cell membranes, tamper with DNA, and even kill cells. This kind of neutralization is called scavenging. Flavonoids became famous after a 1991 segment of the television news show *60 Minutes* examined what was termed "the French Paradox." It was reported that although the French eat a lot of fatty foods such as cheese and butter, they suffer a relatively low rate of coronary heart disease, because they drink a lot of red wine, which is rich in procyanidins. (The segment did more for the American wine industry than the repeal of Prohibition.) Revisiting the topic in 2008, reporter Morley Safer told his *60 Minutes* audience that the explanation for the French Paradox is actually an anti-oxidant in red wine called resveratrol. Safer said that we might all soon be able to take a resveratrol pill that would allow us to live up to a decade longer. This turned out to be journalistic enthusiasm, not science. No one has proven the existence of health benefits derived from either procyanidin or resveratrol, which fell under a cloud in 2012 when the University of Connecticut charged a resveratrol researcher at that institution with falsifying data he had published in scientific journals. Scientists also posit that moderate alcohol consumption alone, or something else in the French diet or lifestyle, might account for the French Paradox, which some critics believe is itself an illusion created by the way France's medical community tabulates disease.[25]

The U.S. Food and Drug Administration has approved neither procyanidin or resveratrol for medical use. It classifies them not as drugs but as "dietary supplements," a wide array of products taken by half of all Americans—although that percentage might fall due to report in a prominent medical journal in 2013 that multivitamins are worthless and a waste of money. Sales of resveratrol—which is classed as a non-herbal dietary supplement even though it is extracted from plants—reached $30 million in the United States in 2012, despite the fact that it has never been tested in human clinical trials. And it may turn out that the only thing benefiting from the belief in the therapeutic value of red wine is California's economy.[26]

In 2012 Americans spent almost $1.5 million on hawthorn capsules, pills, and tinctures processed from the tree's leaves, flowers, and fruit. Of the forty top-selling herbal dietary supplements in the U.S. mass market, hawthorn was listed at number 20 (*Crataegus* is a tree, not an herb, but it is grouped with such plants as St. John's Wort for the sake of marketing convenience). Its sales had increased almost 12 percent from 2011. Cranberry preparations came in at number 1, followed by garlic at number 2, which together accounted for $100 million in retail sales.[27]

Manufacturers of hawthorn products do not need FDA approval because the agency does not classify herbal supplements as drugs and therefore does not regulate them. But it does have the power to require makers to ensure that their products contain what their labels say they do. If not, the supplement can be removed from the market. The agency demands that these products be safe and that they are produced using guidelines called "good manufacturing practices." To that end, in fiscal year 2012 the FDA carried out surprise inspections on 361 firms and issued citations to 70 percent of them. But according to FDA spokeswoman Tamara Ward, "When you look at the numbers of warning letters issued a year and compare it with the numbers of supplements currently on the market, it's safe to conclude that most companies comply with FDA regulations."[28]

There are scores of companies in America selling hawthorn capsules, pills, and tinctures, each product different from the others in terms of content and dosage recommendations. But all are alike in one respect: their manufacturers are not allowed to claim that these products have any specific medical benefits. Vitacost, for example, sells a product called Hawthorn Extracts. These consist of gelatin capsules containing two substances: four hundred milligrams of "Hawthorn Extract" taken from the leaves of "*Crataegus oxycantha*," which has been standardized so that each capsule contains 2 percent "flavonoids"; and six hundred milligrams of extract taken from the hawthorn's "aerial" (any part of the tree that can be seen, such as leaves, flowers, twigs, and fruit). This second component has been standardized to contain 1.8 percent vitexin, one of several flavonoids that have been isolated in *Crataegus*.[29] Although the company acknowledges on the label that the "daily value"—the amount the government determines a healthy American must take of a substance in order to satisfy daily nutritional need—of hawthorn has not been established, it suggests taking two capsules daily "with food or as directed by a physician." And it advises diabetics, people with low blood sugar, pregnant or lactating women, and those with "known medical conditions" or taking drugs to consult with a doctor or a pharmacist before taking any dietary supplement. (To split hairs, the name *Crataegus oxycantha* was rejected by the International Botanical Congress because it was being used to describe several species; the hawthorn in these pills is probably the midland hawthorn, *C. laevigata.*)

Wonder Laboratories also markets something it calls Hawthorn Extract, which also contains two ingredients. First is "Premium Wildcrafted Hawthorn Berry Powder," standardized to contain three milligrams of vitexin. The second is an extract processed from the flowers and leaves. The company claims that this product "promotes heart health" and warns pregnant or nursing women to talk to their doctors before taking it. The recommended dose is one 150-milligram capsule taken three times a day.

In 1982 a vestige of Eclectic medicine made its appearance in Portland, Oregon, when a pair of naturopathic physicians founded the Eclectic Institute to conduct research into botanical preparations introduced by John Uri Lloyd. Among the institute's commercial offerings are a number of *Crataegus* products, including a blend of hawthorn; *Cactus grandifloris,* a native of Jamaica that is also thought to be a slow-acting heart tonic; *Ginkgo biloba,* used to aid circulation; and passion flower, *Passiflora,* believed to be useful against insomnia and anxiety. What is unique about the institute's manufacturing process is that the plants are freeze-dried and reduced to a powder the consumer mixes with water and drinks. A spokesperson for the institute told me that the hawthorn berries, leaves, and flowers used in its products are "wildcrafted" in Oregon from *C. monogyna.* The term simply means that wild plants are harvested in their natural habitat employing methods that do no lasting harm to the plant or its environment. However, when applied to *C. monogyana, wildcrafting* is a misnomer because the species is not native to the United States. As noted earlier, some organizations in Oregon consider it an invasive plant. Still, better that the institute take from the invader than the invaded.[30]

A single clinical trial in France in 2004 indicated that a commercial hawthorn capsule called Sympathyl was more effective than a placebo in treating mild to moderate anxiety disorders. To help screen the 264 patients accepted for the three-month study, researchers used a diagnostic tool called the Hamilton Anxiety Rating Scale. This is a series of fourteen questions intended to rate the severity of unease that people experience, from subjective feelings such as "anticipation of the worst" and "inability to relax" to physical symptoms such as dry mouth, ringing ears, and insomnia. The ingredients in the product, which is manufactured in France, are extracts of hawthorn and California poppy, and magnesium. Although one study concluded that the trial showed that Sympathyl had some effect, the journal of the American Academy of Family Physicians concluded in 2007 that the single study did not prove anything.[31]

A more esoteric hawthorn product from France intended to reduce anxiety is a liquid extract called Élixir d'Aubépine (one of the French names for hawthorn is *épine noble,* "noble thorn," alluding to the belief that the crown of thorns was woven from hawthorn twigs). Made by a company called Élixirs de Provence, it consists of hawthorn leaves, fruit, and blossoms macerated in alcohol, just as in the preparation of other extracts. But in addition, the plant's parts are calcinated (heated until they turn to ash), and the ash is added to the extract in the belief that it contains minerals alcohol cannot capture. This practice, called spagyrics, is rooted in alchemy, which Paracelsus insisted was intended not for the vulgar process of making gold, but for the production of medicine. (The word *spagyric* is from the Greek, meaning to "separate and combine.")[32]

Another growing segment of the U.S. hawthorn market is the production of supplements for dogs, cats, and horses. For $16.99 as of this writing you can buy a one-ounce, nonalcoholic tincture of *Crataegus* that its manufacturer, Animals Apawthecary (*sic*), claims is composed of wildcrafted hawthorn, ginkgo biloba, and garlic in distilled water. The company suggests that the tincture's benefits include "circulatory support" and "antioxidant and strengthening activities." The recommended dosage is ten to twenty drops once or twice a day for cats and small dogs, and twice that for medium and large dogs. Using hawthorn to treat horses got a marketing boost from a scene in the 2003 Oscar-winning film *Seabiscuit,* which was based on a true story. In the film, after stumbling and rupturing a suspensory ligament in his front leg, the legendary horse is nursed back to health by a jockey named Red Pollard. When asked what he has put on the horse's bandage Pollard says, "Oh, that's hawthorn root. It increases circulation." Now *Crataegus* preparations to treat all sorts of equine maladies, from hoof conditions such as navicular to arthritis, are available. It is given to strengthen blood flow and the heart and to reduce tissue damage from oxidation. It is even one of the components in a preparation called Equine Performance Detox marketed by a Tennessee company, Animal Element, to clean the intestinal tract.

In the spring at Dark Acres, after Kitty and I turn out our horses for the day, some of their favorite places to visit are the hawthorn groves. Although not much grass grows in these shadowed corridors, the horses find some nice hors d'oeuvres in the tender new leaves that bud until the middle of June. Our five-year-old, a dun gelding named Dagwood, is so fond of hawthorn I briefly wondered whether he might be medicating himself for a condition we were unaware of. But he has a voracious appetite and seems to be amused by the idea of eating trees (he has also stripped all the bark off a pair of hybrid poplars). Watching him, I'm reminded of the horses I saw in County Waterford with their heads buried in hawthorn hedges, nibbling on the tender tips of twigs and leaves that weren't protected by thorns. In a note to an English magazine, a horse owner reported that in the spring his horse would come in from the field with green ears. Wondering what was causing this, the owner followed the animal and discovered that it was thrusting its head into the hedges. Why, he asked, would a horse eat hawthorns? The editors answered that the horse knew what was good for him: "Your horse is not eating Hawthorn because of any vitamin or mineral deficiency, but because he instinctively knows it has great health-giving properties."[33]

After centuries of wishful thinking and even a few scientific studies is it possible to come to a reliable conclusion about the medical value of *Crataegus?* One organization believes that it is. The Cochrane Collaboration, a global network of ten thousand health experts headquartered in Oxford, is named after Archie Cochrane, a Scotsman who abandoned his medical studies to serve in the Spanish Civil War. He spent World War II in a prisoner-of-war camp in North Africa tending to the medical needs of his fellow prisoners, and there he conducted clinical trials to see which of his treatments worked. Following the war he became a medical professor, and began promoting the formation of an organization that could make critical summaries of all the evidence-based research that was beginning to flood the medical journals. Founded in 1993, the Cochrane Collaboration sifts through published studies, taking into account diet and lifestyle of the subjects and the value of screening for disease. Clinical trials with small samples are given less weight then those with more patients, and badly designed trials are thrown out. The result of its analysis constitutes a systematic review, as opposed to a collection of anecdotes.

In 2008 the Cochrane heart group released a critical assessment of high-quality studies testing the value of hawthorn in treating patients suffering from Class I to Class II chronic heart failure. "Hawthorn extract (made from the dried leaves, flowers and fruits of the hawthorn bush) may be used as an oral treatment option for chronic heart failure. In this review, 14 double-blind, placebo controlled randomised clinical trials (RCTs) were found. They did not all measure the same outcomes and several did not explain what other heart failure treatments patients were receiving. Those trials that could be included in a meta-analysis showed improvements in heart failure symptoms and in the function of the heart. The results, therefore, are suggestive of a benefit from hawthorn extract used in addition to conventional treatments for chronic heart failure." And it concluded that bad reactions to the extract—nausea, dizziness, and complaints of a gastrointestinal and cardiac nature—"were infrequent, mild and transient."[34]

Based on that conclusion I make it a point to eat hawthorn leaves every day in the spring and summer, supplementing them in May with small bites of their fetid, complex blossoms. In the fall I pick haws, mash them with my fingers to force out the seeds, and eat the raw pulp. Sometimes I'll drive down the country lanes around Dark Acres to the sloughs and creeks lined with hawthorns to harvest haws and leaves from trees no one else seems to be interested in, using them to minimize harvest pressure on our home trees. I wash and dry the haws and put them in mason jars with vodka, then store the jars on shelves in our darkroom. In a couple months the concoction turns dark pink. Mixed with a little fresh-squeezed lime juice, it helps brighten Montana's endless winter nights. While I urge readers not to self-medicate with hawthorn (or vodka, or anything else), unless their doctors approve of the practice, I have rather blithely decided that since I'm probably fated to die at the age of seventy-five like my father and his parents before him, I'm going to eat and drink anything I want to that makes me feel better, or allows me to believe I feel better.

In order to have hawthorn on hand all year I use alcohol to make my own hawthorn extract. I wash the leaves, flowers, and mashed fruit pulp and put this in mason jars with cheap vodka, using a ratio of three parts vodka to one part botanical. When the leaves are soft I transfer the mixture to the two-cup jar that comes with a zippy little blender from Nu Wave called the Twister (this isn't an endorsement, it's just what I have on hand). You can buy home-use and commercial extractors for hundreds or even thousands of dollars, but for my purposes, that would be overkill. I run the blender for ten minutes. The centrifugal force and the alcohol solvent open the plant and draw out its essence. Then I pour the viscous, dark-green liquid into a cheese cloth draped over the top of a Mason jar, and squeeze out as much of the vodka as I can. I spread the resulting solid wad onto a cookie sheet and put it in the oven on the lowest setting until it's a dry, crumbly mass, which I smash into a fine powder, and with an inexpensive little capsule-filling device transfer to vegetarian gelatin capsules. How many of these should I take every day? Who knows? Since the dosages in the clinical trials vary, and no one has ever established a minimum daily requirement, I have decided to take three a day with meals.

There are doctors who ridicule old folk remedies. Many are members of a health industry in America that told us alcohol was bad, then said it was good; that blood pressure in people over sixty should be no higher than 140/90, but now it can be 150/90; that eating fat caused heart disease, but now the culprit is sugar. Maybe they're right about herbal treatments. Yet other esteemed voices in the medical community suggest that hawthorn has cardiovascular benefits. The University of Maryland Medical Center, for example, cites hawthorn's reputed anti-oxidant qualities and its improvement of heart function, although the tree has known adverse interactions with synthetic drugs, the center advises, and should be taken with these substances only under a doctor's supervision.[35]

In *Pharmageddon,* the Irish psychiatrist David Healy's lacerating condemnation of the pharmaceutical industry, the author relentlessly cites the deaths and disabilities that have been wrought by synthetic drugs and the zeal of doctors to prescribe them. From the horrors of thalidomide, suicides and birth defects caused by anti-depressants, and most recently the epidemic of deaths and addictions caused by prescription pain-killers, the medical establishment, he argues, too often seems to be influenced by big drug companies whose only interest is profit. A 2013 investigation revealed that scores of Chicago-area doctors were paid to speak on behalf of drug companies. Topping the list was a pulmonologist who was paid more than $160,000 in 2012 to make speeches plugging three manufacturers. While such payments are legal, one company, GlaxoSmithKline, announced in 2013 that it would cease the practice. The company promised it would also stop buying airline tickets and paying hotel bills so doctors could attend conferences, payments which amounted in 2012 to almost $1.25 million.[36]

In a 2010 trial centered at the University of Florida involving 6,400 patients suffering from both diabetes and coronary heart disease, it was shown that hypertension drugs *increased* the risk of death among patients who were taking them to tightly control their blood pressure. The report on the study concluded with grim understatement that this pharmaceutical regimen "was not associated with improved cardiovascular outcomes compared with usual control."[37]

I'll continue to avoid doctors unless I tear a ligament playing tennis or break my arm falling off a horse. And I'll still forage in our hawthorn groves. Hawthorn extract is not just inexpensive and at the worst a placebo, it's also a good source of roughage and vitamin C and may even be life-giving. At any rate, I like the timelessness of foraging for wild plants. It makes me feel a kinship with Dark Acres that I wouldn't feel if I merely sat on the patio and watched the trees sway in the wind.

NINE

A Tree for All Seasons

More grass means less forest; more forest less grass. But either-or is a construction more deeply woven into our culture than into nature, where even antagonists depend on one another and the liveliest places are the edges, the in-betweens or both-ands. . . . Relations are what matter most, and the health of the cultivated turns on the health of the wild.

—MICHAEL POLLAN

On bright summer mornings there's so much birdsong streaming from the canopy at Dark Acres the place sounds like a tropical rain forest. This stretch of timber, grassland, slough, and shore we share with another couple got its name because by late afternoon it's cast in deep shadows thrown by the Bitterroot Range on the left bank of the Clark Fork. Towering ponderosa pines and black cottonwoods define the overstory; the understory is dominated by Douglas hawthorns, willows, water birch, and red osier dogwood. Only twenty-three acres in area, in volume this enormous aviary is twice the size of the old Houston Astrodome, a sanctuary in a spreading exurban landscape, never logged or farmed.

While there are 165 species of birds native to our region, I'd become curious about how many individual birds were actually in attendance here on a typical day. Certainly dozens, maybe scores. Could there be hundreds? I had assumed a census would be impossible because birds come and go as they please. But I learned that there *is* a way to count them—two ways, actually—if you have the patience and the expertise. The first method is the fixed radius point count. Several trained observers place themselves in the centers of imaginary circles with a radius of, say, a hundred yards, and write down all the birds inside the circles they see and hear for a fixed period of time, usually five or ten minutes. Then, often using GPS devices to navigate, they move to other spots in the centers of circles whose circumferences touch that of the first circles. And so on. This method yields a list of species present on a finite piece of ground on a given day, and a reasonably accurate estimate of their numbers. Of course, the more observers the more reliable the numbers.

Just after dawn on an August morning, volunteers from the Five Valleys Audubon Society arrived at Dark Acres. Led by Jim Brown, a retired Forest Service specialist in wildland fires, the four researchers soon set off into the floodplain using a deviceless survey called the transect method. Walking together as a foursome for three hours along our game trails, they compiled sightings and soundings to produce a ballpark inventory of the kinds of birds at Dark Acres and their numbers. On this morning (mornings are the best time to count birds because they're active: hungry and not yet affected by the rising heat), they counted thirty-seven species, ranging from a Sandhill crane and a pair of bald eagles to a pileated woodpecker and a gray catbird. The birds they were most excited about were the nine Lewis's woodpeckers they saw. After their survey, as we drank coffee on the back patio, we watched these showy red, pink, and iridescent green flycatchers launch themselves from the leafless upper reaches of a cottonwood, snatch insects from the air, and return to their perches. Brown explained that Lewis's woodpeckers are considered a "species of concern" because the loss of its habitat in Missoula County to urban development has begun to reduce the bird's numbers.

In all, Brown and his crew counted 178 individuals. He said that during breeding season, from May to July, when some unseen birds make themselves known by singing, the Audubon Society would have counted many more and identified a larger number of species, so I promised to invite them back in the spring. The next morning I went out by myself and saw some birds that hadn't been at Dark Acres the day before. These included a trio of magpies, some red-winged blackbirds, and a Western meadowlark. I heard, or thought I heard, a raucous Steller's jay.

Although most of the species of birds at Dark Acres tend to avoid our hawthorns and their thorny embrace, the American robin, cedar waxwing, Bohemian waxwing, black-capped chickadee, pine grosbeak, and black-headed grosbeak have co-evolved a relationship with the tree that's based on its fruit, which isn't on the menus of most birds here because it contains only about 2 percent fat. When the haws ripen in late July, haw lovers mob the branches. But they don't eat everything. The reason for this restraint is unknown, but maybe it's an old instinct compelling the bird to put aside something for leaner times. In the winter, haws that have survived the warm-weather feeding frenzies still hang on the trees, desiccated, but a critical source of cold-weather nutrition.

While most robins migrate south in the winter, a few winter at Dark Acres because of the lingering haws and remaining berries on other plants, such as the snowberry and the western juniper. If all the robins stayed up north there would not be enough of food to last the winter. But I rarely see these occasional homebodies in the winter because they hide in the hawthorn thickets and don't emerge to hunt for insects until spring. The waxwings and the chickadees never migrate. The chickadees winter inside holes gouged into the soft stumps of dead cottonwoods, lining their nests with moss and rabbit fur. On cold winter nights they lower their body temperatures, a form of torpor rare in birds. In the fall their diets change from insects to berries and seeds, which they hide in huge numbers. One study suggests that in October the bird sheds old neurons in its tiny but powerful brain and grows new ones imprinted with the exact location of all these hiding places.[1]

A few birds harvest the fruit in the summer and fall, but then either head south for the winter or turn to other foods. For ruffed grouse the berries are only a small part of an enormously varied diet, from watercress and dragonflies to salamanders and snakes. The shelter, more than the fruit, attracts grouse to the hawthorn thickets. In the spring we know when a male bird is looking for a mate because we can hear him beating his wings, a courtship ritual known as "drumming."

Our three species of thrush—the veery, the hermit thrush, and Swainson's thrush—are shy birds that bulk up on haws in the fall. But what they really like about our hawthorns is the dense shade inside the thickets, where they can hide from predators as they forage for bugs. All three species migrate to the tropics for the winter, usually flying at night. Members of the flocks keep in touch with one other using a distinctive contact call.

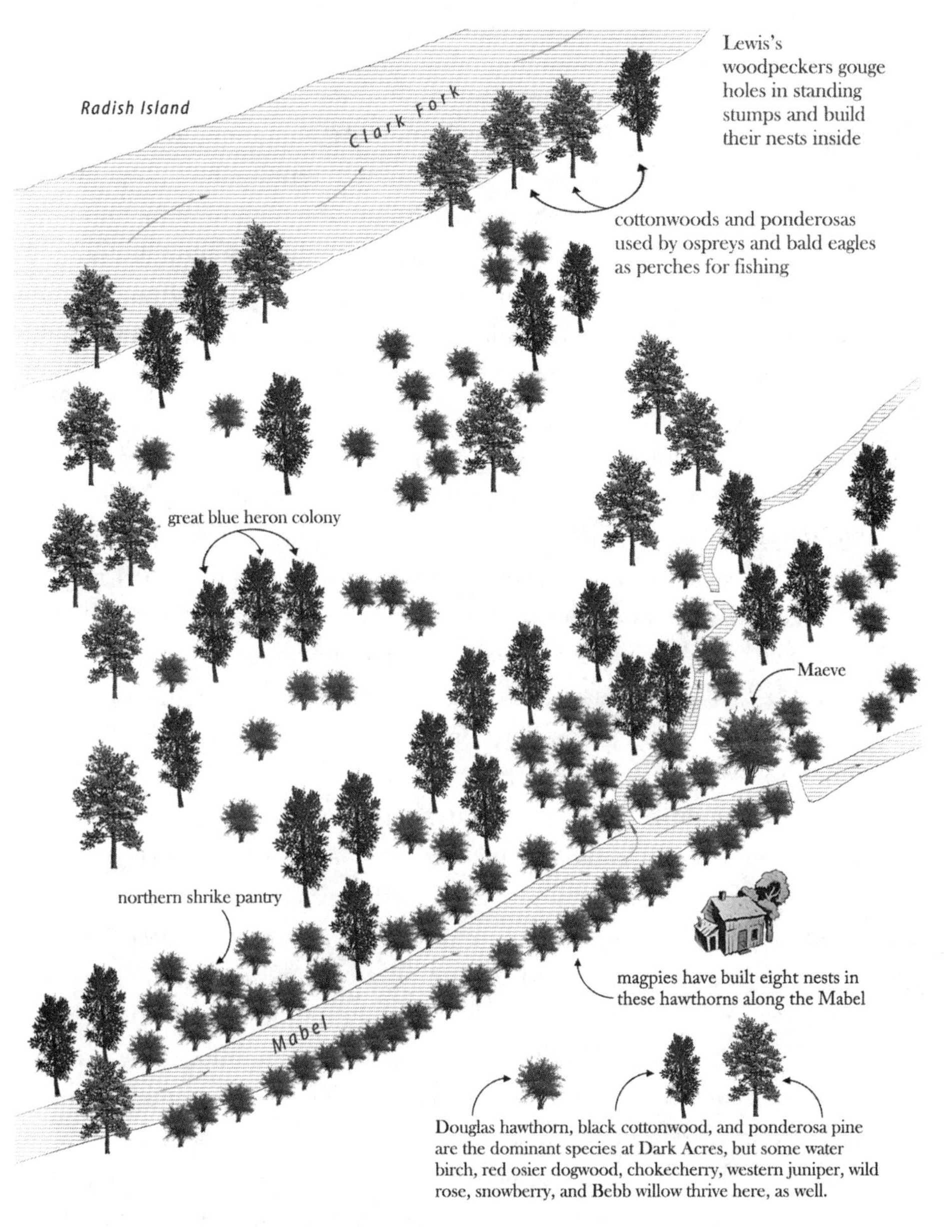

The relative distribution of trees at Dark Acres (illustration by the author)

In exchange for its fruit, birds transport the hawthorn's seeds to distances that range from a couple of yards to more than half a mile. They are the tree's best agents of reproduction. Inside a *C. douglasii* haw are three to five small seeds called nutlets. After the birds eat the seeds but before they excrete them, gastric enzymes and hydrochloric acid in their digestive tract reduce the hard skin that covers the nutlets. Although this process is unnecessary for the reproduction of the tree, it speeds up the germination time. The coating, called an endocarp, is why the fruit is classified as a drupe, although it resembles a pome, the type of fruit produced by *Crataegus* cousins in the Rosaceae family such as apples and pears. While the nutlets contain cyanide in order to repel insects, the concentration is low, and the nutlets pass through an animal's digestive system fast enough to prevent acids from dissolving the endocarp.

A few bird species prefer to raise their chicks in our hawthorns to take advantage of the protection from predators and the elements offered by the tree's spikes and dense foliage. Beginning in March, black-billed magpies, *Pica hudsonia,* build huge nests high up in the tallest thorns. First the female scoops mud from the banks of our sloughs and builds a cup-shaped base lined with grass in the crotch of a sturdy limb. The male, which is heavier and slightly larger than the female, delivers load after load of dead hawthorn twigs and small branches to the work-site, and both birds fashion this lumber into a sloppy, loosely stacked structure rising from the cup. Then they cover the nest with a domed roof. They typically install two entrances, one on each side, possibly to provide an escape route in the unlikely event of a predator braving the thorns. Because it can take as long as seven weeks to build a new nest, magpies often claim unoccupied nests, instead, and make the necessary repairs. Once the female lays her eggs, usually six or seven, which are pale green with gray flecks, the birds become fiercely territorial. But they are interested only in defending their own personal hawthorn, one reason why among the ten mature *C. douglasii* lining the right bank of the Mabel, our widest slough and a prestigious neighborhood for magpies, there are eight nests. The biggest of these is more than three feet high.[2]

In the spring Kitty and I like to grab binoculars and watch the magpie show. This is a constant and theatrical parade of birds bringing food to their ravenous chicks—insects, mice, nuts, berries, eggs, carrion, and garbage. They also hide treats to eat later by plunging their beaks into the ground, stuffing the food into the hole, and covering it with twigs or leaves. If they think another magpie is watching they'll move the food. When they are satisfied with their hidey-hole they cock their heads and stare at the cache, apparently memorizing its location. In the sun their

A magpie and its nest, engraving by Harrison Weir, from Mary Howitt, *Birds and Their Nests* (London: Partridge, 1885), 105

black-and-white plumage becomes iridescent, reflecting colors that in turn flash green and blue with hints of red and yellow, like a rainbow. Cranky and exuberant, the magpies yell at their neighbors, producing an array of warbles, squawks, clicks, and coos that make a single bird sound like a small flock. Magpies regularly mimic the calls of other species: It's not unusual to hear a meow or the barking of a small dog, while townspeople have reported hearing what sound like sirens. A member of the Corvid family that includes crows, the black-billed magpie can be taught in captivity to mimic the human voice.[3]

Magpies often engage in behavior that seems odd compared to that of other birds. When our horses graze near the slough, magpies sometimes fly down and strut around on their backs, scavenging for ticks and looking for cuts from which they can extract a blood or meat snack. The horses seem to enjoy it (at least they tolerate it). When a black-and-white magpie is riding our black-and-white paint, Rolex, the two resemble Me and Mini-Me, from the Austin Powers movies, or a circus act. Some stockmen dislike magpies because this behavior irritates saddle sores, shearing cuts, and branding wounds.

Magpies also engage in a peculiar behavior known as "anting," in which the bird snatches up an ant and rubs it on its body. Occasionally a magpie will lie down on a colony and squish the ants into its feathers by rolling around on them. Sometimes the bird will simply alight on a hill of ants, stir them into a frenzy and let them swarm all over it. No one knows why magpies do this. But there are theories. One of these holds that anting releases the formic acid found in the venom of ants (the Latin word for "ant" is *formica*). Formic acid is a counterirritant that acts against bacteria, fungi, ticks, and mites, pests that plague birds. Another theory holds that formic acid augments the preen oil that waterproofs a bird's feathers, which is secreted by the uropygial gland. Still another says that the magpie is removing the distasteful formic acid from the ant before eating it. My favorite theory concludes that to a magpie an ant massage feels so good the practice becomes addictive.[4]

A 2008 study suggests that magpies possess something that is extremely rare in the animal world—the power of self-recognition. Researchers placed a small colored adhesive dot on the necks of five domesticated magpies in a spot where the birds could not see it. Then they were put into a compartment containing a large mirror. Two of the birds studied the image they saw in the mirror, then reached up to their necks with a claw and snatched away the dot. Anyone who has seen a robin attack its own image in a plate-glass window will understand the significance of the act. Besides humans, only the great apes appear to recognize themselves in such a way. Of animals that have been given the mirror test the magpie has the highest percentage of brain weight to body weight.[5]

One overcast autumn day while I was looking for firewood I came upon a congregation of magpies gathered around something on the ground. I crouched down to watch, figuring they were brunching on carrion, perhaps the remains of a deer that coyotes or a mountain lion had killed. But when one of them flew off I could see that the object of this meeting was a lifeless magpie. The cause of death was not visible. The birds were taking turns gently pecking at the corpse. Then a magpie flew in with a beak full of leaves and laid them on the fallen bird. Another magpie flew off, returned with leaves, and repeated the gesture. Marc Bekoff, professor emeritus of ecology and evolutionary biology at the University of Colorado, published a report on a similar ceremony, concluding that it was nothing less than a funeral, a way for the living to say good-bye to the dead. He later received e-mails from people who had also witnessed magpies and other Corvids practicing these last rites.[6]

A bird that's a big fan of the black-billed magpie is the long-eared owl, *Asio otus.* Instead of building its own nest it moves into the abandoned nests of magpies. Ranging in weight from half a pound to a pound of mostly feathers, with the females sizably heavier than the males, long-eared owls hunt at night for voles and the occasional gopher or mouse, which they can pinpoint in complete darkness because of their exceptional auditory skills. One reason for this acuteness is the asymmetrical arrangement of its ears—the left ear opening is higher than the right. When the owl hears a rodent, it swoops down, kills the rodent with a bite to the back of its head, and usually swallows it whole. Animal parts that it can't digest, such as hair and bone, are regurgitated in pellets popularly known as puke balls that fall to the forest floor, marking the bird's presence. Studies of these pellets have shown that 95 percent contain the skull of a vole—something we have at Dark Acres in abundance.[7] The only time we hear the low hoots of these birds is during their mating season, from March to June, and we rarely see them. The owl's streaked, brownish plumage blends seamlessly into bark, and it can make itself even less conspicuous by compressing its feathers. We might be looking directly at one of these owls and not see it.

Because long-eared owls are active only at night, for a long time no one knew exactly what they did in the dark. But in the spring of 2013 the wildlife biologist Denver Holt, with the Owl Research Institute in Montana, installed an infrared camera and a microphone above an owl's nest north of Dark Acres. Holt told me he was surprised to learn that when they are raising their young, owls rarely sleep. They might nod off, but their eyes spring open at the slightest sound. And they made noises no one had ever recorded before. As I watched the owl cam, which

A long-eared owl and its prey (Phant/Shutterstock)

was being broadcast online, a pair of magpies showed up at the nest during the midnight hour to cock their heads at the owls. Maybe they were anticipating a meal, maybe they were angry because the owls had taken over their nest, or maybe they were just curious.[8] During the previous winter Holt and an associate had walked into a scrubby hawthorn thicket on a creek a couple of miles from Dark Acres to flush long-eared owls into a series of nets so they could put bands on the birds and track their movements. (It had been a good season for owls because the vole population in western Montana was higher than it had been recorded in many years.) "This little patch [of hawthorns] is real typical of habitats you don't think are very important," Holt explained in a video recording the visit. "But for long-eared owls this is a very important winter habitat and sometimes breeding habitat. This is what we call a truck stop because it has such a high turnover of owls, whereas the other areas around Missoula, once the winter roosts are established, usually in October, are pretty much stable throughout the winter."[9]

While some of the birds that love our hawthorns use them for critical but prosaic needs, two species of shrike have evolved a more exotic relationship with the tree. The northern shrike, *Lanius excubitor,* and the loggerhead shrike, *Lanius ludovicoanus,* are the only North American songbirds that consistently prey on vertebrates. Although they differ slightly in appearance and range, both species are gray and white and slightly smaller than an American robin, *Turdus migratorius.* They hunt by taking up position high in the canopy. As soon as they see a victim below them they drop down on it and knock the bird to the ground with a blow from their bills, which are hooked like a falcon's and have a sort of tooth on the upper jaw that fits into a gap on the lower. Sometimes they grab the victim with their feet, wrestle it to the ground, and bite it. Shrikes have to use their bills for this purpose because they don't have talons. They also attack voles and mice, plus an array of insects and small reptiles and amphibians. Sometimes they consume their prey at once, but they might also impale their victims on the spikes of hawthorns, other thorny plants, or even barbed wire fences. (Occasionally, the victims are still alive.) During lean times the birds will return to this pantry for their meals. Because they kill more than they can eat at one time, shrikes were once called "butcher birds" and a group of shrikes is still known as an "abattoir."[10]

According to one study, the northern shrike uses food to buy sex. When mating, a male brings the female to his larder to show off his prowess as a hunter. Apparently, the larger the size of the impaled prey, the more likely the female is to mate with him: a nice, fat mouse makes a better showing than a couple of bugs. A female will choose a male with a well-stocked pantry, according to this theory, because while she is incubating the eggs and nurturing the chicks he will have to provide the food. But male shrikes also seek extrapair couplings. A male might offer food to another male's mate—in fact, males will offer a new female better food than he offers his own mate (rather like a husband who buys chocolates for his wife but mink for his mistress). The higher in energy the morsel the better his chances. Extrapair copulations offer males the opportunity to fertilize more eggs and females more free food; presumably they both enjoy the sex.[11]

On 2 January 2000 some western Montana birders were thrilled to witness the first recorded appearance in the state of the vermillion flycatcher, *Pyrocephalus rubinus,* a small, brilliant-red bird that for some reason had flown thirteen hundred miles from its natural range in the

A shrike impaling its pantry of prey on thorns, from Jules Trousset, *Nouveau dictionnaire encyclopédique universal illustré,* 6 vols. (1886–91) (Morphant/Shutterstock)

Southwest. One day in the Bitterroot Valley a group of admirers waited three hours to see the celebrity and were rewarded when it landed near them beside a pond. It gorged happily on insects for a few minutes, then rose into the air. Suddenly, to their horror, a northern shrike rose into the air as well, and began chasing the exotic visitor. The shrike slammed into the smaller bird a few times, and they disappeared behind a house. The last thing the birders saw of predator and prey was the shrike flapping away with the flycatcher in its beak, which it transferred in midair to its claws.[12]

The first harbinger of spring at Dark Acres is a tiny yellow buttercup. It usually opens in mid-March, a sudden pinpoint of vivid color in an otherwise drab carpet of dead grass that's been crushed under snow and ice since November. In mid-April the leaf buds of our *C. douglasii,* the Douglas hawthorn, start to open. But not until the trees blossom—which happened on 13 May for three years in a row—do I know that warm weather is here to stay. The hawthorn's unfolding of white begins at the crown and works its way down the foliage. A week after it begins all the trees in a vast corridor of thickets along the drainage of the Columbia River, from deep inside Montana seven hundred miles to the Pacific, are in full bloom, filling the senses with their fetid perfume. Lying in the shade of these trees watching me, our newest stock dogs, acrobatic little Border collies named Hanna and Zoe, suddenly lift their noses to the sky and jump up to search for this wonderful dead thing so they can roll in it.

Heading to the river, we cross over a culvert on the Mabel, which rises and falls with the seasonal flows of Clark Fork, three hundred yards west. A trio of western painted turtles dive from their log at our approach, and a muskrat ducks under the water to hide. Across the slough, in the tallest of our hawthorns, an eastern fox squirrel is eating the last of the dried haws that have been hanging there all winter. Much to the dismay of people who are protective of our two species of native squirrels, *Sciurus niger* has spread across western Montana after being introduced by a Midwestern doctor who enjoyed these rodents and brought some with him to Missoula in the 1960s. I'm trying to rid Dark Acres of plant species that aren't Montanans because they tend to force out native plants. I mow and pull and burn the spotted knapweed and leafy spurge that have invaded the place (knapweed poisons defenseless natives by infecting the soil with a chemical called catechin). And I began putting a chainsaw to our Russian olive trees. But while another invasive species that dines in the thickets is the raccoon, I no longer have the stomach to pull a trigger. One July a big raccoon lost its quarrel with a bald eagle and fell from a cottonwood, landed on his feet, hissed when he saw me staring, and fled into the hawthorn groves.

In the confusion of hawthorns on the left bank of the Mabel a white-tailed doe is nibbling the lower leaves. In the fall she'll reach up to eat the haws as well. Other animals that rely on this fruit include several species of vole and the field mouse. In the fall, black bears sit on their haunches or rise up on their back legs to get at the dark purple haws, which they shovel into their mouths with both paws. Although there are no grizzlies at Dark Acres, observers have reported them thirty miles north of here doing the same thing. I've never witnessed a black bear climbing a hawthorn, but the videos I've seen of it make me wonder how they avoid getting pricked.

Today I've decided to mark this explosion of spring by counting our hawthorns. Although only eleven of the twenty-three acres belong to us, I consider our neighbor's place part of Dark Acres because it has exactly the same kinds of habitat. The land upstream and down from these two parcels has been so abused and overgrazed the only trees that survive are a few junipers and ponderosa pines, making Dark Acres and the adjoining spread into a sort of woody island. Counting *C. douglasii* in riparian corridors where it's abundant, is fuzzy math, at best. The tree not only tends to grow multiple trunks, it also shoots up a lot of suckers—trunks growing from buds in its roots that can pop up some distance from the original trunk. When I finished counting several hours later I added up my column of figures. Only then did it occur to me that maybe the 136 individual trees I just counted were actually only two, one on the right back of the Mabel and one on the left.

Very little direct sunlight penetrates the thickest of the groves because of the tangle of dead and living branches, some drooping down to touch the ground. But this heavy year-round canopy provides critical cover for birds and animals, a place to escape from predators. And the hawthorns help keep them warm by blocking the wind. The few plants that grow in these somber and scary thickets have developed a relationship with the hawthorn only one side of which I understand. Wild raspberry and wild rose, hawthorn cousins in the Rosaceae family, become aggressively vinelike once they latch onto a hawthorn and climb as high as twenty feet along it toward the sun. The western clematis, *Clematis ligusticifolia,* is a true vine that produces delicate white blossoms. European species of clematis were paired with hawthorns in England to encircle Elizabethan arbors where lovers came to embrace, though we have one of these vines at Dark Acres that is decidedly unromantic. Eighty feet long and two inches in diameter, it coils under its *Crataegus* host at one end like an enormous snake. In early July, clematis vines shroud our tallest hawthorn with white blossoms, making it appear as if the tree is blooming once again. All this effusive fruiting and foliage and flowering evoke the image of Flora in Botticelli's strange masterpiece, *Primavera.* Although they make a pretty sight, it's a mystery to me what the tree gets out of serving as a trellis for various climbers.

You might assume that hawthorns have no choice about what plants climb them. But like a lot of species of trees, they are believed to have the ability to poison their neighbors with organic toxins called allelochemicals. The study of allelopathy is fairly new, and controversial. Researchers theorize that trees attack other trees by disrupting the microbial life of the soil, altering the cycle of nitrogen fixation and the permeability of root membranes, which limits a plant's ability to suck up nutrients and water. Some scientists maintain that obvious acts of aggression are not due to chemistry at all but to the effect of competition for sunlight, water, and soil. Because the sun does not shine in a thicket few plants can grow there. But other scientists maintain that *C. douglasii* ranks somewhere in the middle on a continuum of floral assassins, exuding what is thought to be a "moderate" amount of allelochemicals. One piece of evidence supporting the hawthorn's negative effect on other trees is the existence of fields in the Midwest recovering from farming that so far have been populated only by hawthorns.[13]

One of the most potent arboreal chemical weapons is exuded by the eastern black walnut *Juglans nigra,* a North American native valued for its timber. It releases small amounts of a phytotoxin called juglone from its roots, leaves, and bark that inhibits the growth of everything from mountain laurel and lilacs to cabbages and tomatoes, though one genus that is not affected by juglone is *Crataegus.* (What the allelopathic effect of hawthorns might be on black walnuts is unknown.) A study of eucalyptus in California, where several species were introduced from Australia, revealed a "bare zone" around a plantation of these trees. A number of chemicals, such as highly toxic terpenes and phenolic acids, were found in the tissues.[14]

Twelve miles upstream from Dark Acres is a battlefield where Douglas hawthorns have been waging what was until recently a losing war against Norway maples, *Acer platanoides.* Greenough Park in Missoula is a half-mile-long corridor of urban riparian forest on the banks of a mountain stream called Rattlesnake Creek. Norway maples are an invasive tree introduced to the colonies from Europe and probably brought to Missoula in the 1860s by the city's founder, Frank Worden, from his home in Vermont. After most of America's chestnuts were killed by a blight introduced from Asia, towns like Missoula began planting Norway maples along streets because they are attractive and tolerate pollution (as do most species of hawthorn, but hawthorns aren't touted as street ornaments because they smell funny and have thorns). But wherever these maples are allowed to flourish nothing else grows. The University of Montana wildlife biologist Erick Greene describes what they created in Greenough Park as a "wall of death." They are among the tallest trees in the park, rising more than sixty feet, and when they stand close together their dense canopy casts everything below them into shadow. Their roots grow close to the surface of the ground, stealing moisture from other plants. Some scientists also believe that they secrete allelochemicals that attack the competition. Because the maple is a foreigner, it lacks natural enemies in America, where the natives have not had enough time to evolve a defense against it. Neither grass nor even most weeds can grow under the maple's crown. For other trees life around these maples is a struggle. Although only one *C. douglasii* has died since we moved to Dark Acres—killed by a falling cottonwood limb—scores of Greenough's Douglas hawthorns have been killed by these invaders. They held their own for a while, possibly producing their own chemical weapons, but the maples finally grew tall enough to cast them in their shadow.[15]

Greene presented a report to the city council in the 1990s predicting that within two decades Greenough would become a Norway maple monoculture. So the city began cutting them down. Where the maples were cleared the natives have returned, if slowly at first (some researchers suggest that the trees' allelochemicals linger in the soil). On a walk through the park one day I came upon a palisade of towering maples on one side of the path. The ground under them was bare and their canopy was silent. On the other side of the path, where the maples had been cleared, new Douglas hawthorns and black cottonwoods had grown shoulder high, and were hosting a noisy party of birds.[16]

While *Crataegus* has been used for eons by people across the northern hemisphere as a source of medicine, the trees have trouble fending off their own diseases. Almost every one of the *C. douglasii* at Dark

Acres becomes infected in the summer with an unsightly fungus called cedar-hawthorn rust. In May, when the leaves have fleshed out, they are bright green on the bottom and darker green on the side turned toward the sun. Two to four inches long, leathery and waxy to the touch, they are slightly serrated and shaped like snowshoes. A month later, after the trees have blossomed and are growing tiny green haws, the first spots will appear on the tops of the leaves. A red dot surrounded by a border of yellow, these are an eighth to a quarter of an inch in diameter. As the summer passes, a tiny bulging sac will emerge from the spot on the underside of the leaf, and little white antennae called spore horns sprout from it. As the haws mature, some of them become infested with the antennae, as well.

The fungus, from the genus *Gymnosporangium,* leads a very strange life, which is saying something for a fungus. First, borne by the wind, its two-celled spores light on junipers, forming hard, pea-sized bumps called galls, which sprout spore horns. After the first warm spring rain the spore horns become gelatinous and turn orange. The spores released from the spore horns fly on the wind and infect the hawthorns. In July spores from the hawthorn fly back on the wind to the junipers, where they lodge in crevices along the twigs. There they germinate, completing and then beginning again the circle of fungus life, which requires both trees as hosts. The disease infects most species of *Crataegus,* although some, such as *C. phaenopyrum,* are resistant. I have seen the fungus on hawthorns all over western Montana and on the *C. viridis* I visited at Chanticleer in Pennsylvania. But I didn't happen to see it on any of the *C. monogyna* I looked at in Ireland.[17]

There are two other fungal infections that cause occasional problems for hawthorns, although the trees at Dark Acres have never shown any symptoms of them. One is powdery mildew, which produces whitish splotches that cause leaves to curl up and die. The other is apple scab, a common malady on the west coast that ruins fruit and distorts leaves. These diseases can be controlled by making sure the tree has access to full sunlight without any irrigation. And in the fall it helps to rake up all the leaf litter these fungi like to spend their winters in, and burn it. Copper-based fungicides are effective weapons, as are some organic agents such as neem oil.

A disease of seventy trees and bushes in the Rosaceae family, fire blight, earned its common name from the fact that it causes fruit and foliage to wither and turn black, as if scorched with a blowtorch. Sometimes a honey-colored slime oozes from infected sores. When the malady reaches the roots the plant dies. The culprit is the bacterium *Erwinia amylovora,* which is carried by everything from rain and wind to insects and birds. It enters the tree's body through natural openings, or lesions caused by bugs and damaging weather events such as hailstorms. North American hawthorns like *C. douglasii* have evolved considerable resistance to the bacterium because it is also thought to be a native. But for orchard growers the disease is a serious threat to their livelihoods, especially those who cultivate pears and apples. It was first discovered in the Hudson Valley in 1780 and moved west as virtually every frontier homestead was planted with European varieties of apples to meet the huge demand for alcoholic cider. An 1844 epidemic in the Midwest wiped out many orchards. In 1905 it was reported in Montana. It appeared in England in 1957 and crossed the English Channel in 1972, infecting *C. monogyna* growing in Belgium and the hedgerow country of Normandy, where native hawthorns have no resistance to the disease. It jumped from the hedges and attacked cider apples thirteen years later. To combat it, orchard owners began destroying infected trees and planting varieties that showed an ability to fight off the infection. Streptomycin has been extremely effective for the past half-century at wiping out the bacteria, although strains resistant to the antibiotic were reported in 2011 in New York, sending a shudder through the fruit industry.[18]

While our *C. douglasii* have their health issues, they do better than some of the other trees at Dark Acres when it comes to defending themselves against plundering insects. For example, some of our younger ponderosa pines have been killed by the mountain pine beetle, *Dendroctonus ponderosae,* which lays its eggs under the bark after injecting the tree with a fungus called blue stain that halts the production of resin, which would kill the bug or force it out of the tree. Hawthorns are immune to this pest. But there are lots of other predators that victimize the tree. Still, our *C. douglasii* don't seem to be bothered by some insects that attack other species of hawthorns, such as leafminers, hawthorn lace bugs, and the borers that ruined many nineteenth century hawthorn hedges in the mid-Atlantic states. In the eastern United States the eastern tent caterpillar moth, *Malacosoma americanum,* finds the leaves of hawthorns and other members of the rose family a very tasty meal and can damage trees if too many of them congregate. Insecticides are useless against these caterpillars, but all sorts of creatures like to eat them.

Rhagoletis pomonella is a tiny fly less than a quarter-inch long that employs an evolutionary survival strategy called Batesian mimicry, named for Henry Walter Bates, who studied the phenomenon in Amazonian rainforests during the mid-nineteenth century. The fly's otherwise transparent wings are marked with a dark shape that looks like an "F." When it feels threatened *Rhagoletis* extends its wings at a 90-degree angle from its body and moves them up and down while walking sideways. The new position of the F on its wings presumably gives predators such as spiders the impression that this isn't a fly, it's a spider, and not worth fighting with over territory. The adults, who communicate almost entirely through the release of pheromones, mate on the fruit of hawthorns, and the female lays her eggs just under the skin, one at a time. The larvae hatch ten days later and begin eating tunnels through the yellowish, mealy flesh of the haw. The adults eat the skin. After the Puritans brought domesticated apples to America from Europe, some flies abandoned the hawthorns and took up with the exciting new fruit, bigger, sweeter, and juicier than the bland little haws. This preference earned *Rhagoletis pomonella* its name, "apple maggot." While most news about nature focuses on the extinction of species, the apple maggot, which has destroyed entire orchards, is on the verge of doing something more interesting: evolving into two separate species, one whose genetics is based on apples and another that is still associated with *Crataegus.* Driving this divergence is the fact that flies meeting and reproducing around apples never come across flies that live in hawthorns because of the order in which the fruit appears.[19]

Other *C. douglasii* pests include a couple of species of aphids, soft, defenseless, pear-shaped bugs that are some of the 4,400 species of Aphididae despised the world over because they destroy crops by sucking out a plant's juices and sometimes injecting toxins and harmful viruses. Under the right conditions they multiply with a bizarre zeal. They are not only born pregnant; their embryos contain embryos, like

Russian nesting dolls. The only fun thing about them is the peculiar relationship some species have with what are known as dairy ants. Aphids convert the sap they extract from leaves into an amber-colored fecal matter called honeydew, which the dairy ants like to eat, as do other insects, birds, and animals (including people). The sugar-loaded liquid is sometimes produced in such abundance it drips from plants where aphids are feeding. This happens every summer at Dark Acres, when the *Nearctaphis sclerosa* aphids infest our hawthorns. Aphids eat so voraciously they cannot get rid of their honeydew fast enough so they rely on the dairy ants to "milk" them. The aphid raises its abdomen to signal that it's milking time, the ant strokes the abdomen, and out comes a highly nutritious snack. In exchange, the ants protect the aphids from predators, and sometimes invite them to live in their nest. When winter comes they might herd their aphid cows down into the burrow for safekeeping, returning them to the hawthorn in the spring. Sometimes the ants get carried away and eat the aphids. But this usually happens when there are too many aphids to milk.[20]

One visitor to our hawthorns that loves aphid burgers is the flower fly maggot, *Eristalis tenax,* which looks more like a small caterpillar than like those creepy little white things that live in putrid meat. The adult is often mistaken for a bee or a wasp because of its black-and-yellow stripes and its habit of pushing its abdomen into things as if it were stinging, although it doesn't have a stinger (another example of Batesian mimicry). By impersonating a dangerous insect the harmless flower fly frightens away creatures that like to eat flies. Also known as the hover fly, it moves like a helicopter, darting sideways and pausing over a hawthorn flower before diving down to drink its nectar. While the fly is feeding, yellow pollen attaches to the fly's body and is transported to the next flower.

Flower fly maggots are sometimes joined in their aphid feeding frenzy on our hawthorns by lady beetles (the Coccinellidae family) and the larvae of the lacewing (the Chrysopidae family).The adult lacewings, which eat nectar, pollen, and honeydew, are fragile-looking insects with enormous transparent wings. Their hearing is so acute they can pick up the ultrasound calls of bats, which like to eat them. Lacewing larvae resemble little alligators. They sprout long bristles that collect dirt and garbage, which helps camouflage them against ants that are guarding their aphids. Because lacewings are such voracious predators they are sometimes called aphid lions. They inject a venom that paralyzes their prey, slice them open with their sickle-shaped mandibles, and suck out their juices. In the three weeks before it transforms into an adult a single larva might eat as many as six hundred aphids or other destructive, soft-bodied pests such as spider mites, mealybugs, and whiteflies. Lacewings are used for biological pest control in place of insecticides, which kill both helpful and harmful insects. One company, Arbico Organics in Arizona, sells lacewing eggs for about ten dollars per thousand, with large discounts offered to volume buyers. Because lacewing larvae tend to become cannibalistic in confined quarters, they must be shipped overnight. This is also why the mother lacewing lays her tiny white eggs on individual quarter-inch stalks: to prevent the newborns from eating one another.

Another biological way to control aphids is by means of the *Aphidius* wasp. Blackish and a mere eighth of an inch long, the female wasp lays a single egg inside an aphid, where it hatches. The larva and the pupa spend their life cycle completely inside the aphid, feeding on its innards. The aphid becomes bloated and discolored, and eventually dies and becomes mummified. Finally, the adult wasp chews its way out and emerges into the light of day. Another kind of wasp, *Lygocerus,* is a hyperparasite that lays its eggs only in aphids that already have an *Aphidius* growing inside them. When the *Aphidius* grub has eaten the aphid, the *Lygocerus* grub eats the *Aphidius.* Several companies specializing in the biological control of insect pests sell bottles of aphid mummies containing pupa that are ready to emerge. These are useful in controlled environments such as greenhouses, but in the wild the wasps are unreliable, often flying away from the plants they are intended to police.

In our back yard, a few yards from the Douglas hawthorns lining the right bank of the Mabel, is an old Siberian elm, *Ulmus pumila,* planted years before we moved to Dark Acres. In the summer of 2013 it became so infested with the Siberian elm aphid, *Tinocallis saltans,* that the sidewalk underneath it was constantly splattered with honeydew, around which swarmed wasps, flies, and myriad other bugs. One explanation for this infestation was the weather—it was warm enough at the beginning of summer for aphids to hatch, but too cool for lacewings. When it got really hot and dry the aphids already had the upper hand. Well, maybe. Later the same summer the hawthorns were plagued by aphids, as usual, but they didn't rain down honeydew. This may have to do with the fact that our hawthorns are natives, a keystone species providing livelihoods for a balanced community of native insects, and the elm is an exotic alien that upsets the natural cycle. Besides its constant shedding of dead branches and its lack of vigor, its foreignness is another reason why it's on my list of things to cut down.

Our hawthorn flowers attract a menagerie of pollinators. One of these is the smallest breeding bird in North America, the gorgeous calliope hummingbird, *Stellula calliope.* The male attracts females and discourages competitors by elevating the feathers in the magenta-and-red gorget around its throat, producing a starburst effect. It laps nectar with its long extendable tongue. Several species of bees are fond of this nectar, as well, including the western honeybee, *Apis mellifera.* As it eats, the bee picks up grains of the yellowish pollen on its coat. It places most of these in hair-fringed structures on its hind legs called pollen baskets. Then it mixes nectar and saliva with the pollen to make food, "bee bread," for its larvae. The pollen that remains on its coat rubs off on the next flower, fertilizing it.

Sweat bees (the family Halictidae) are beautiful, metallic-colored insects that take their name from their attraction to the salt in human perspiration. We also have mining bees (the genus *Andrena*).The female digs a single tunnel that branches into other tunnels that terminate at chambers called brood cells, which she packs with a wad of nectar and pollen. She lays one egg in each of these nurseries and seals the chambers. This catacomb is designed to protect her young from marauding ants. When the larvae hatch they're greeted with a feast. Some bees visit the same trees over and over during blossoming, a behavior called trap-lining. Because they are unable to see the color red but can see ultraviolet, they prefer white flowers such as those of our *C. douglasii* to flowers of other colors. The honey produced from most species of *Crataegus* tends to be dark amber in color, wild-tasting and thicker than, say, honey made from clover.

Some bugs are attracted to hawthorns because they savor the smell of death. These are members of the order *Diptera,* the true flies, which have two wings as opposed to, say, bees, which have four. The chemical responsible for the odor is trimethylamine, an organic compound that is released as a gas in May by the blossoms of our *C. douglasii.* Variously described as smelling like ammonia or rotten fish or semen, it is the same substance formed in the first stages of decomposition in human and animal flesh. For that reason some of the flies that pollinate hawthorns are also the first to arrive at the scene of a crime. Their presence helps forensic entomologists figure out how long the body has been dead and whether it was moved. (I sometimes wonder whether the hover flies, blow flies, and common houseflies I watch drinking nectar in the hawthorn blossoms have been rooting around recently in a nearby cadaver—a deer, maybe, or one of the dogs or people who disappear from time to time while swimming in the picturesque but dangerous Clark Fork.)

Blowflies (the family Calliphoridae) often arrive at a corpse within minutes of death. The females lay their eggs in body openings such as the nose or the vagina or a wound, then fly away. The eggs hatch within two days. The maggots develop in three larval stages called instars before entering a pupa stage and then emerging as immature adults. Investigators can establish time of death by determining which stage a bug is in and noting its size. If an investigator finds a blowfly with crumpled, undeveloped wings near the body, for example, it's likely that the person has been dead for sixteen days. Common house flies, *Musca domestica,* can also tell secrets about a cadaver. Fourteen days after they lay their eggs in a corpse, mature flies emerge and hang around for awhile bringing new maggots into the world. (A fun fact about house flies is that their feeding parts are designed for drinking, not eating: the house fly vomits saliva and digestive juices on a piece of sugar, meat, or feces in order to turn the solid into a liquid, which it sucks through its straw-like tongue.)

Caterpillars eat flowering plants such as hawthorn, which is on the menu of more than fifty kinds of butterfly and moth larvae worldwide. Two of these Old World bugs are the larvae of the hawthorn shield bug, *Acanthosoma haemorrhoidale,* and the hawthorn moth, *Scythropia crataegella,* a leaf miner that dines almost exclusively on *Crataegus.* The adult version of the hawthorn shield, a stink bug that looks like a Stealth bomber, can grow up to three-quarters of an inch long and is colored red and green to camouflage it in Irish and British hawthorn trees bearing ripe fruit. Like the house fly, it liquidizes leaf tissue with an enzyme in its saliva, then drinks the beverage through a tube. It extracts fruit sugars from ripe berries in the same way. When it is upset it excretes a smelly orange liquid that deters predators.

The *C. douglasii* at Dark Acres host three species of butterfly, providing leaves and twigs on which the insects lay their eggs. These are the striped hairstreak, *Satyrium liparops,* the gray hairstreak, *Strymon melinus,* and the mourning cloak, *Nymphalis antiopa.* After its eggs are deposited, the striped hairstreak spends its days lounging in the highest branches of the thickets, soaking up the sunlight needed to power its flights down to sip the nectar of flowering weeds and sap leaking from trees. The mourning cloak, which is Montana's state butterfly, is drab gray on its underside, with an extravagant, iridescent palette of dark maroon on the top of its wings bordered by a band of cream, inside of which is a line of cobalt blue spots. This striking coloration apparently resembles the garment that grieving Germanic people wore to honor the dead. When its wings are folded it is almost impossible to see it against the shredding gray bark of the *C. douglasii,* whose leaves are eaten by the caterpillars from the moment they hatch. The mourning cloak can live up to a year, which is a long time for a butterfly. It does not migrate, as many butterflies do, but hibernates in the shelter of loose bark. At Dark Acres, we've often been surprised to see one emerge while there's still snow on the ground. When it's startled, it emits an audible click as it flies away, apparently intended to frighten predators. The males are intensely territorial and spend much of the mating season menacing one another, as well as birds. An English entomologist and broadcaster, L. Hugh Newman, fell in love with this gorgeous insect, known as the Camerberwell beauty in Britain, where it is not a native. Hoping to populate England with them, he raised thousands on his butterfly farm in Kent, marking them with dye so they could be identified later. In 1956 he released them into parks. The next spring not a single one could be found.[21]

The pleasure I got from watching the narcotic dance of butterflies was muted when I learned that although they get most of their nutrients from nectar, they will also feed on human cadavers. Attracted by the smell, which they detect through their antennae, the butterfly will land on a corpse and "taste" it with organs in its feet. If it likes what it tastes it will use its proboscis to suck up the sodium-rich compounds that have seeped to the skin during decomposition, a process known as puddling. Because almost all puddlers are male, scientists believe that the salts and minerals end up in the sperm and are transferred to the female's eggs. Fortifying the eggs with supplements gives butterflies a better chance of hatching robust caterpillars that will mature into robust butterflies and presumably pass on their parents' genes. Butterflies will puddle on carcasses, rotting fruit, and feces, but they most commonly puddle on, well, mud puddles, which contain the sodium the insect craves. While nectar is loaded with sugar it doesn't have all the nutrients the insect needs for successful reproduction.[22]

The wood-products industry in the Pacific Northwest has been fading for years. In its heyday it built communities all across the region, including Missoula and Eugene, Oregon, which were once rough logging camps but have now been gentrified into university towns. For me, the most obvious symbol of this industrial decline sits in ruin down Mullan Road a couple of miles north of Dark Acres. Before it closed in 2010, it was a factory, a vast complex that manufactured one item, the brown kraft paper used to fabricate corrugated boxes. For fifty years a parade of logging trucks would enter the mill with pine logs, and trainloads bearing huge rolls of paper would leave it. Production was a round-the-clock operation, a roaring, clanging, hissing bedlam that created its own weather and injected the overcast night skies with colors not seen in nature. Now the plant looks like it has been bombed: it's being torn apart for recycled steel. There are groves of *C. douglasii* surrounding it, but the paper industry has no interest in them: one *can* make paper out of hawthorn, but the process of harvesting it, then reducing its adamant mass to fine sheets with pulpers and chemicals, is commercially impractical. The hawthorn is too skinny and gnarled to be of much use as timber. About the only things crafted from its fine-grained wood, which sands and polishes to a beautiful sheen, are boutique items such as handles, decorative boxes, and inlay.

But in fact Douglas hawthorn is the dominant species of the dense understory below much taller trees stretching from western Montana throughout the Columbia River Basin, south to the Bay Area and north to the Alaska panhandle. Before the Columbia rose behind its dams you could have walked from Butte, Montana, to the Pacific without losing sight of a hawthorn. The central role the tree plays in the environment of the region is most apparent in May, when it's in full bloom, and corridors of white blossoms follow the rivers and creeks as far as the eye can see. I don't know of anyone who has made the calculation, but it's possible that the tree produces more fruit than all the orchards in its range put together. If we consider other places in the world where hawthorns constitute a sizable percentage of the biomass—Tasmania, China, Britain, Ireland, and other parts of Europe—a good question might be what kind of role will it play in the coming crisis wrought by global warming. Because hawthorns are often the first trees in the northern hemisphere to populate abused ground where a decimated forest is destined to return, should we be planting more of them in these places? (One such reforestation is taking place in the Upper Peninsula of Michigan, where the U.S. Forest Service is trying to restore an isolated population of native *C. douglasii* in nine counties where it has been reduced by human interference.)[23]

One winter afternoon I wandered through our hawthorn groves looking for tracks. I was curious to see if I could catalogue all the different birds and animals that were taking refuge from the season under these domes of gnarled branches. Where did this strange tree come from, I wondered, in form so different from the soft, straight conifers and cottonwoods that tower above the floodplain? And really, what's the meaning of all these vicious thorns?

TEN

Essence and Spinescence

It is not the strongest or the most intelligent who will survive but those who can best manage change.

—CHARLES DARWIN

The hawthorn may have come from another world. Some scientists posit that the beginning of life on earth 3.7 billion years ago might have been the continuation of life that began on another planet. Mars is most often cited by enthusiasts for this theory, which holds that primitive organisms thriving on what was then the watery red planet were catapulted to earth inside meteorites during the Late Heavy Bombardment of the solar system, when asteroids smashed into the inner planets with such force that hunks of planet were knocked into space, and only spores or simple organisms capable of hibernating would have survived the journey. According to another bombardment theory, this rain of meteorites deposited organic chemicals and water inside the craters they gouged. Heated by rock melted during the collisions, which also cracked open thermal vents, these depressions became crucibles that cooked a primordial soup in which chains of organic molecules were formed. When the chains began to react with one another, setting off a series of chemical reactions that lasted millions of years, the result was a metabolizing sludge that grew and spread across the planet, laying down deposits that eventually became the world's supply of petroleum. According to some researchers, the first life on earth arose from this sludge, which existed in a state between alive and inert.[1]

Our *C. douglasii* can trace their family tree back to insects living down the road from Dark Acres in the hot springs of Yellowstone Park. One of them, *Thermus aquaticus,* thrives in saltwater heated to 160 degrees Fahrenheit or more inside subterranean chambers surrounded by magma. Shaped like a worm, this anaerobic bacterium will die instantly if exposed to free oxygen. The conditions in the thermal pools are probably very much like those on earth when it was covered in sludge. Scientists believe that the volcanic atmosphere then was a mixture of ammonia, carbon dioxide, and water vapor, and the surface of the planet was many times hotter than it is today, a paradise for oxygen-hating microbes. Researchers were astounded in the late 1970s when a "new" form of life was discovered in Yellowstone's hot springs, something that absorbed sulfur and produced methane. These one-celled creatures were classified as Archaea, and they are so unlike any other life form that they were placed in a third kingdom of life between bacteria on one side and complicated organisms such as human beings and stink bugs on the other. Since their discovery, Archaea have been found thriving in a variety of environments: under ocean beds, in petroleum deposits, in the digestive tracts of fish, cows, and termites, and in acid mine wastes.[2]

At some point in the organism's ancient history a wandering bacterium entered an Archaeum, and what eventually evolved was a hybrid whose cells resemble those of hawthorns and humans more than those of bacteria and Archaea. These cells were constructed of discrete parts, organelles (little organs), which are the remnants of the bacterial passengers, each designed to perform a specific job. While the DNA in bacteria cells floats around, a hawthorn's DNA is neatly packaged inside the membranes of a nucleus, and respiration is carried out by an organelle called a chloroplast. This is the remnant of a cyanobacterium, a microbe that developed the ability to synthesize light into energy. Because its waste product was oxygen, scientists suggest that while transforming the earth's atmosphere it drove the anaerobic bacteria to the brink of extinction. The chloroplast is responsible for photosynthesis, in which sunlight is converted into carbohydrates, which the cell burns for fuel. About a billion years ago the direct ancestors of all plants appeared. This was green algae, a multi-cell organism that lived exclusively in water. Its appearance concluded the heavy-lifting stage of evolution. Everything that followed was merely a variation on a theme.

Half a billion years ago some green algae washed ashore or were driven there by a storm. These accidental tourists found wet places on land and thrived, although they no longer had the ocean to buoy and transport them. Some algae evolved into primitive plants such as mosses. While mosses have been wildly successful and are still found everywhere, they never evolved a way to grow upright and remained close to the earth. They are unable to store water so they tend to dry up in droughts and spring back to life when it rains. Mosses are not the ancestors of trees: they have no roots, leaves, vessels, or structures that function as backbones to give them the strength to rise up toward the sun. One of the first plants to evolve a trait central to the evolution of the tree was a genus called *Cooksonia,* which lived in swamps. *Cooksonia* consisted of branched stalks a couple of inches tall with saucer-like heads bearing spores that scattered in the wind. Now extinct, they had a rudimentary vascular system in which water stored under pressure allowed the organism to stand upright. Modern plants that rely on the same force, which are classified as herbs, include banana "trees" and the eight-foot cattails that share a slough-side terrace at Dark Acres with some of our hawthorns.[3]

One of the most peculiar life forms from this era was what scientists who studied its fossils have dubbed a "humongous fungus." As wide as three feet in diameter and standing as tall as twenty-five feet, *Prototaxites* were branchless spires that narrowed at the top to a blunt point. During their prime they towered over every other living thing on earth.[4] In terms of biomass, another humongous fungus is the largest life form on the planet today. Sprawling across three and a half square miles of Oregon's Malheur Forest is a single organism, the *Armillaria solidipes,* also known as the shoestring fungus. A parasite living in the soil and in the roots of conifers, which it kills and then feasts on, this fungus is thought to be between 1,900 and more than 8,600 years old.[5] In the same area are four other individuals, smaller but still deadly to conifers. Every fall after the first rain, this malevolent family sprouts millions of what are called honey mushrooms. Although slimy and not as tasty as the morels that pop up in May at Dark Acres, they're edible. The occasional Douglas hawthorn that grows in the same area, which is higher in elevation than the parts of Oregon where hawthorns flourish in huge groves, is resistant to *Armillaria.* In fact, the hawthorns may be profiting from the misfortune of the conifers because of the destruction of their canopy, which brings the smaller trees more sunlight.

After the evolution of *Cooksonia* came a period of phenomenal generation resulting in strange, swampy forests teeming with enormous ferns, horsetails, and a genus of treelike *Archaeopteris,* for a long time regarded as the world's first tree. But in 2007 scientists excavating in a sandstone quarry in Gilboa, New York, announced that they had unearthed a fossil of an earlier tree. Resembling a long-handled brush, this tree flourished 380 million years ago, when Gilboa was located near the equator in a tropical swamp on the shores of an ocean terrorized by armored fish weighing four tons. Because the quarry was about to be excavated to provide stone for highway work, two paleontologists racing against the deadline used an industrial saw and an engine hoist to lift the fossilized crown of the tree from the earth. Given a reprieve by the quarry owner, they went back and excavated the trunk in pieces, which they assembled like a jigsaw puzzle.[6]

While *Wattieza,* this spore-bearing genus, *looked* like a tree, it was as different from one of our hawthorns as a Venus flytrap is from a buttercup. It had roots, leaves, a trunk, and a reproductive system, but the resemblance to a tree ends there. For one thing, there is nothing in the fossil evidence to suggest that *Wattieza* had a tap root. Instead, it sprouted a number of small tendrils that apparently acted as anchors. Like many hawthorn species, *C. douglasii* sends down a long tap root in its first years, and then develops a shallow but extensive root system that produces suckers.

Wattieza's enormous leaves resembled the wispy, droopy fronds of ferns, whereas *C. douglasii's* leaves are a couple of inches long and seem flimsy until you touch one. (If you study the veined and wrinkled architecture of these little electrical generators, you can understand how the leaves of trees inspired the new generation of flexible solar cells made of plastic, which were invented in 2011.) What made *Wattieza* more treelike than fernlike, however, was its trunk. Although only five or six inches in diameter at the top and twenty inches at the base, the tree stood more than twenty-six feet tall. It was able to achieve this height for two reasons: as with modern trees, the cells in its trunk were packed together in tight sequences and bound to one another with an organic super-glue called lignin, which reinforced the cellulose in the cells. (On their own these cells exist merely as limp filaments—the cotton that rains down from our black cottonwoods every July is cellulose and nothing more.) Vascular plants without lignin in their stems droop during droughts and then desiccate. But a thirsty tree can stand firm until it drinks again because its fibers are tough and hard. (Paper and rayon are simply realigned mats of cellulose fibers whose glue has been removed.)

As in *Wattieza,* the core of a modern tree is made of heartwood, which despite its name is formerly living material surrounded by live sapwood that contains the vessels that transport nutrients. Sapwood is ringed by cambium, tissue that contains stem cells that direct the growth of sapwood on the inside and bark on the outer. Although it is not possible to confirm this, the wood of hawthorns is probably many times harder than *Wattieza*'s was. In general, the wood of slow-growing trees such as most hawthorns is denser and more adamant than that of fast-growing trees like weeping willow.

How hard is hawthorn wood? The standard test of hardness in wood is the Janka test, developed in 1906 by Gabriel Janka, an Austrian researcher, and refined over the years. A steel ball is driven into a plank of cured heartwood that is "clear" (free of knots) until the ball reaches a depth in the wood equal to its radius. The force required to do this produces a number—the higher the number the harder the wood. The U.S. Forest Service's Forest Products Laboratory in Madison, Wisconsin, offers a list of some 260 species of hardwoods that have been tested. (One of the laboratory's notable achievements was determining that the reason an abnormal number of Major League baseball bats were breaking in 2008 was because of the alignment of the grain in the maple used to make them.)[7]

But hawthorns are not on the lab's Janka list. Laboratory botanist Michael Wiemann told me that this was because *Crataegus* is not considered a commercial wood. The Janka test is extremely useful for the manufacturers of furniture and flooring because it indicates how well their products will resist breaking and abrasions. Hawthorn wood, as noted earlier, is generally used only for decorative purposes because it is so thin and gnarled. Wiemann also speculated that when Janka testing was performed a sample of hawthorn might have gone under the ball, but like some species of hickory, which are also not included in the results, hawthorn tends to split when dry, so it would not have been given a number.[8] (I briefly considered cutting down one of our *C. douglasii* and finding someone to test it. In order to avoid the problem of splitting I could get a sample of green heartwood. While Janka samples are air-dried to a standard moisture content of 12 percent before they're tested, there is a formula that would yield a number for hawthorn that I could use to compare it with other hardwoods. However, because I have exchanged the Catholicism of my childhood for another set of superstitions, killing a hawthorn is now something I would never risk.)

An alternative method for gauging hardness is by means of specific gravity. This is a measure of density determined by the amount of water a sample of green heartwood displaces. The figure is then divided by the weight of the same piece of wood after it dries. To find out the specific gravity of hawthorn, I didn't have to cut down any of my *C. douglasii* because the specific gravity of several *Crataegus* species had already been measured by persons who apparently were not worried about the consequences of committing Crataegucide. Although *C. douglasii* was not one of the hawthorns tested, because the numbers assigned to these species fall into a narrow range I'll assume that our home species is not much different from the hawthorns that were, and conclude that its specific gravity is .78.[9] This number means nothing until it is compared with the specific gravity of other kinds of trees. The hardest wood in the world is lignum vitae, which comes from a small, slow-growing Caribbean tree. Its specific gravity is 1.26, and even when it is completely dry it will *sink* in water. One of the hardest North American woods is Osage orange, whose specific gravity is .85. Hawthorn is harder than the sugar maple used to make baseball bats (.63), the red maple flooring in our house (.49), our black cottonwoods (.30), and the softest wood in the world, balsa (.18), as well as most of the other six hundred species of tree in North America.

Wattieza was asexual. It reproduced by releasing one-celled spores; if these landed in a nurturing place, they would turn into a multi-celled organism—that is, another *Wattieza.* If its reproduction functioned like that of modern ferny plants, its spores were held in an enclosure, a sporangium, on the end of a nifty little device called an annulus that covered the undersides of the fronds. When the annulus unfurled, in a forceful two-step motion, the spores were catapulted into the wind. The advantage of this sort of reproduction is that unlike sperm and eggs, spores can remain viable for a long time while they wait for rain or some other form of moisture. One disadvantage of spore reproduction is that the offspring are all clones of the parent, resulting in a lack of diversity that can limit the ability of a species to respond to its environment and adapt in order to survive.

Hawthorns have evolved three breeding strategies. One is conventional sexual pollination. Pollen is germinated on the stamens in the flowers of one tree and deposited by insects, birds, or wind on the stigmas in the flowers of another tree, where it travels down a canal called the style to the ovary and fertilizes the eggs. These then develop into fertile seeds surrounded by pulpy fruit.

The individuals of some species can also be fertilized with their own pollen. But many species of hawthorn are capable of bringing forth a new generation through the a process of virgin birth called agamospermy, a form of apomixis occurring in some flowering plants that produces offspring without the union of male and female parts. One day the ovaries begin to swell, and soon the tree manufactures fertile seeds called nutlets surrounded by fruit just waiting for a robin to carry new hawthorns far and wide. These trees will be clones of their mother, genetically identical in every way. (Weirdly enough, however, some virgin births rely on the contact of pollen with the ovaries to get the ball rolling, although the pollen itself doesn't accomplish the fertilization.) From an evolutionary point of view apomixis offers advantages and disadvantages. Although it allows some hawthorns to mass reproduce like a Buick assembly line, the lack of diversity resulting from generation after generation of clones might eventually lead to a dead end. However, apomictic trees avoid extinction by reverting to sexual reproduction from time to time. Virgin birthing may have evolved as a way for the tree to reproduce in the absence of pollinators or cross-pollinating male partners.

While apomixis gives some *Crataegus* an advantage over other plants, it is one of the reasons botanists have had nightmares trying to identify the genus by species. Part of the problem stems from the work of early American taxonomists. One such was Charles Sprague Sargent, who identified hundreds of new "species." Born into a wealthy Massachusetts family in 1841, Sargent graduated from Harvard, served in the Civil War, and returned home to the family's 130-acre estate to work as its self-taught horticulturist, a new occupation in America. Instead of the artificial landscape of geometric gardens and topiary that was the current rage, Sargent created a more natural landscape of wild woods, winding lanes, and groves of twelve-foot rhododendrons. In 1872 he was appointed director of Harvard's new Arnold Arboretum in Boston, and between 1882 and 1888 he published a lavish and influential work about trees, *The Silva of North America.* The trouble for subsequent students of hawthorns began when Sargent and two other botanical workers, William Willard Ashe and Chauncey Beadle, turned their attention to *Crataegus.* Before 1896, taxonomists had described about one hundred North American species. But by 1925 the list had been inflated to nearly eleven hundred. Sargent was responsible for more than half these additions. He based his identifications on small, arcane differences in flower, leaf, bark, and haw. It took decades for botanists to determine that most of Sargent's "species" were not different species but slight variations within a species. Decades later botanists described the situation he created as "a veritable witches brew" and a "taxonomic disaster."[10]

To begin with, Sargent did not understand apomixis, especially the fact that when a hawthorn clones itself it often passes down subtle mutations from one generation to the next. An ambitious taxonomist might read these as evidence of a new species, maybe one he could name after himself. Finding himself in a vast grove of *Crataegus* whose individuals appeared to be related, but manifested small differences, Sargent could easily have become carried away in announcing new finds. He also did not realize that sexual hawthorns are promiscuous and breed with other species of *Crataegus* to produce hybrids. The Douglas hawthorns at Dark Acres are "self-compatible"—they can fertilize themselves with their own pollen. They are also apomictic. Their ten stamens are used as one marker to differentiate them from a similar species called *C. suksdorfii* (Suksdorf's hawthorn), which has twenty stamens, and is named after a pioneering botanist in the Pacific Northwest, Wilhelm Nikolaus Suksdorf.

Hawthorn reproduction can be assisted by humans as well. It is possible to grow a hawthorn from a cutting, though it is challenging because it takes many weeks to and conditions have to be perfect—but no one is sure what "perfect" conditions are. It is easier to graft a cutting from one species of hawthorn onto the body of another, or even onto pear, medlar or quince (and vice versa), which are all cousins in the Rosaceae family. *Crataegus* can flourish in poor soils that its cousins dislike, providing them with a robust rootstock they can use to take water and nutrients as the sapwoods of the two species grow together. This makes it possible to plant fruit in more places. And it makes it possible to change the kind of fruit being grown more quickly than replanting an entire orchard and waiting for the trees to grow. Grafters are careful to use a hardy, native hawthorn that is the product of many generations thriving in the same neighborhood. Adventures in grafting have produced such marvels as one-tree orchards—a single hawthorn bearing pears, quinces, and medlars along with several other species of hawthorn. Quinces are pearlike pomes that must be cooked before they're edible. Medlars are delicious, but they aren't edible until they're "bletted," or browned by frost, so they can be harvested later than other products of the orchard. There are only two species of medlar. One was cultivated as far back as three thousand years ago in Turkey, where it is native. The other is a rare and endangered species discovered in Arkansas. Originally believed to be a kind of hawthorn, it was determined in 1990 that Stern's medlar was actually a hybrid produced by the union of medlar and hawthorn, which has thorns and hawthorn-like foliage when it is young, and medlar-like fruit when it is mature.[11] How it ended up in Arkansas, no one knows.

Besides the novelty and challenge of giving life to a Frankenstein-like collection of parts, why would anyone want to graft a hawthorn onto another hawthorn? One good reason is that it produces fruit much more quickly. A growing number of orchardists in the South are pairing cultivars of the mayhaw, *C. aestivalis,* with hawthorn rootstock. (A cultivar is a version of a plant that has been selected for some desirable quality, then bred over and over to enhance it. In the case of mayhaws those qualities are size and taste.) Commercially harvested mayhaws grow up to almost an inch in diameter, are bright red, loaded with pectin—a natural jelling agent—and so tasty they are used in pies, jellies, syrup, fruit leather, and wine. Routinely harvested in the South before the Civil War, mayhaws are making a comeback in an expanding market. They are one of the two varieties of North American hawthorn whose fruit is valued as anything more than an emergency food.[12]

The other prized hawthorn fruit comes from *C. mexicana,* the species native to the highlands of Mexico and Guatemala that the Aztecs called *texoctl* and is still known as *tejocote* in these countries. (Fourteen other *Crataegus* species native to Mexico are also called *tejocote;* their haws range in color from chrome yellow to a copper-orange and can grow as big as two inches in diameter.) Eaten raw, the haws of *C. mexicana* lack flavor, like the pomes of most hawthorns, but when cooked they are transformed into a sweet-and-sour treat reminiscent of plums and apricots. Like mayhaws, the cooked fruit produces a thick syrup because it contains high levels of pectin. Prepared with unrefined cane sugar, cinnamon, and guavas or other fruits, tejocote is the main ingredient in a traditional Mexican hot punch, *ponche Navideño,* served at Christmas and on New Year's Day. On the Day of the Dead celebrations that last from 31 October to 2 November the haws, and candy made with tejocote and chili powder, are offered to the dearly departed. Rosaries made of haws adorn altars. On the pilgrimage taken every 12 October from Guadalajara to honor the Virgin of Zapopan the faithful wear necklaces of strung tejocotes, which infuse the experience with the aroma of roses and apples, and provide snacks during the five-mile walk.

Because the fruit plays such an emblematic role in the holiday experiences of many Mexicans and Mexican-Americans, U.S. customs officials once braced every Christmas for a flood of people trying to smuggle tejocotes across the border into California. They had to be smuggled because they were banned after it was discovered they harbored a fruit fly native to the highlands that could ravage California's fruit industry. Between 2002 and 2006 tejocote led the list of illegal fruits seized by the U.S. Department of Agriculture. In a 1997 raid more than nine thousand pounds of fresh Mexican tejocotes were confiscated in the produce district in downtown Los Angeles and other local markets, where the fruit was fetching eight to ten dollars a pound.[13]

The solution to the smuggling problem was simple: make tejocote available in California, and smugglers would lose their market. But developing a domestic supply was not so simple. After the tree was introduced to California in 1868 some modest attempts were made to cultivate it, most notably by Luther Burbank in the late 1800s at his fifteen-acre experimental farm in Sonoma County. The guiding light of California agriculture, Burbank introduced more than eight hundred varieties of commercially useful plants, including new plums and prunes, the precursor of the Idaho potato, his famous shasta daisy, a thornless blackberry, and a spineless cactus, which is fed to cattle. But some of Burbank's sixty thousand adventures in botany produced freaks, such as when he impregnated a California dewberry, *Rubus ursinus,* with pollen from a hawthorn and a number of other fruits in the rose family. The dewberry sprouted copious flowers and produced its typical raspberry-like fruit, from which he extracted the seeds and sprouted them. The seedlings grew and eventually flowered in a red, white, and pink riot of radically different blossoms. But only two of these five thousand plants produced fruit, and it was ugly and tasted awful. When he extracted the seeds, hoping to produce something more appealing from them, he discovered that they were hollow shells.[14]

In 1914 Burbank wrote that the hawthorn "is an extremely valuable shrub, and gives very great promise of the production of improved varieties of fruit through selective breeding. . . . With the Hawthorn I have made some interesting experiments, but there is fine opportunity for other workers in this field. Indeed, the work of developing this fruit has made only the barest beginnings." Presumably, the hawthorn he was referring to was *C. mexicana:* there is one growing today on the grounds of Burbank's Gold Ridge Farm. But growing right next to it are two hybridized *C. pinnatifida,* a very tasty species native to China that produces bright red, one-inch pomes. Burbank envisioned that the reengineered fruit of his hawthorns would someday be as sought after as any other product of California's prolific orchards. For more than a century no one else agreed.[15]

In the late 1990s, after one of the USDA's tejocote raids, a produce vender asked a Mexican-American orchardist named Jaime Serrato if he would be interested in growing the fruit for the U.S. market. In 2000 Serrato grafted pathogen-free cuttings of *C. mexicana* onto rootstock in groves in San Diego County, where he raises guavas, sweet limes, and other kinds of exotic produce whose importation from Mexico was also forbidden. Four years later he harvested his first crop from a thornless variety of *Crataegus.* By 2010 Serrato Farms was the largest tejocote grower in the United States, selling the harvest produced on thirty-five acres to retailers such as Superior Grocers and Tapia Produce. People who wanted to grow their own trees were buying them from several southern California nurseries. This modest but lucrative market was threatened when the Mexican government petitioned to allow orchardists in Mexico to export fresh tejocotes to the United States that had been irradiated and certified to be pest-free. Most of Mexico's crop comes from commercial orchards supplying retailers who charge less than a dollar a pound. Serrato filed a comment in opposition to the plan, claiming that the fruit fly problem in Mexico is so widespread that as much as 30 percent of the crop is destroyed by the pest and implying that it was not possible to guarantee pathogen-free shipments of the fruit. As of this writing, the USDA has completed its assessment of the petition, but has not yet made a ruling.[16]

Hawthorns are also put to another, more esoteric use. This is bonsai, the venerable art of shaping dwarfed trees, which was introduced by the Chinese to the Japanese and then to the rest of the world. Although it requires patience to shape hawthorns because they are slow-growing trees, they make intriguing subjects for the bonsai artist because their limbs can be readily trained to assume the kinds of aesthetically pleasing forms intended to bring their onlookers inner peace. In 2003 Graham Potter, an English horticulturist, rescued a century-old *C. monogyna* from a hedgerow in Wales that was scheduled for removal as part of a habitat-restoration project, and transplanted it to a traditional stoneware bonsai pot. "Hawthorns really resent root disturbance," he explains in a video documenting the makeover of this little tree. "So my own personal preference is to leave them in the pot for five to ten years." In early spring, before the tree leafs out, it's a homely-looking thing with a short, thick trunk three feet high, lopped off at the top, and a menacing confusion of bare, thorny branches. The emotion it elicits in me isn't serenity, it's apprehension.[17]

In the video, the first step Potter takes in transforming the hawthorn from an unsightly gnome to a work of art is to trim back the long twigs that have sprouted in response to its most recent pruning. Next it has to be disarmed. "If we don't remove the thorns," Potter explains, "we'd just be knee-deep in blood by the time we're finished." Potter then uses a power carving tool called a terrier to reshape the flat, unattractive chainsaw cut at the top of the trunk and reconfigures the branches by wrapping them in pliable, heavy-gauge wire twisted and pulled so as to achieve the overall shape of the crown he wants for the bonsai. Using a wire sling, he draws in a larger branch closer to the trunk. Although these measures seem invasive, the tree responds with a flourish of new growth. "In bonsai we have to encourage the tree to be beautiful," Potter says, "rather than forcing it into submission." He believes that the secret to success has more to do with sound gardening practices than conceits about art.

Two months later the tree has sprouted a mass of leaves and tender new branches. Now is the time, he says, to prune the new twigs, before they become lignified, or stiffened with wood. Hawthorns are different from other bonsai trees such as American maple and Chinese elm in that they respond eagerly to early trimming by producing a profusion of new foliage. In the final step Potter snips off all the vertical branches, called "water shoots," that have sprouted from buds on the old wood. These shoots are desirable in a newly laid hedge, but in a bonsai they need to be removed to give the tree a chance to use its saved energy to produce flowers and fruit. Potter told me that after one full growing season he removes the wire. The branches will by then be set in place, retaining their shape and giving no evidence that wire was used.[18] The result of Potter's lavish attention is a perfect miniature tree whose balance and form suggest a ballerina, blossoms in her hair, arms draped above her head as she dances at the climax of her performance.

A pivotal event in the evolution of the tree was the development of the seed. Spores were, and still are, a successful means for a plant to reproduce because the parent can beget a huge number of them, and they can sprout even after long periods of time spent in unfavorable conditions. During the Carboniferous period, when the world's deposits of coal were created from decaying vegetation, some spore-bearing plants called scale trees grew as tall as 130 feet. But unless the wind carries spores to a place that is or becomes moist and bathed in sunlight they will eventually die. (A spore might be likened to me as a kid, a feral boy running off to spend the day wandering around the floodplain of the Missouri River with no particular direction in mind, looking for fun.) But a seed is equipped with its own food supply of fat, protein, and carbohydrates (me with a full pack on a four-day hike after it was determined that a stint in the Boy Scouts might teach me some discipline). By the time the seed exits the tree its embryo has already developed into something that is ready to become an instant plant. But it will not germinate unless it senses that the conditions are right for success, something that might take years. Hawthorns native to the colder parts of North America will not germinate until they have been on the ground for at least two winters.

When they ripen in late July, each of the three-eighths-inch dark-purple haws produced by our *C. douglasii* contain four or five tiny seeds about the size of the kernels in a bowl of Grape Nuts cereal. In the *C. monogyna* native to Ireland the berries contain only one seed, and in *C. laevigata,* the midland hawthorn, there are two to three seeds. But in any case, these little nutlets are probably not a good thing to swallow. Hawthorn seeds are pyrenes, which are composed of a bit of embryo and plant food protected by the endocarp. This is in contrast to the seeds of quinoa, figs, or sunflowers, for example. Like the pits of other fruits in the rose family, such as peaches and apricots, the seed inside the endocarp of hawthorns contains a small amount of a chemical compound called amygdalin, which produces hydrogen cyanide when it is metabolized by an enzyme in the human small intestine. Amygdalin is also the source of laetrile, the compound that was touted as a cancer cure until coroners identified it as the cause of several deaths from cyanide poisoning, and clinical trials proved that it was useless in treating disease. Although hawthorn endocarps are unlikely to dissolve during their thirty- to forty-hour passage through your digestive system—a third of that time period in bears—there's no good reason to eat them if you're not a bear. Seeds are typically removed from the berries when making food products.

The fabulists who wrote the Old Testament believed that after Eve talked Adam into eating the forbidden fruit God lost his temper and put a curse on his sinful human creations. "Because you listened to your wife and ate from the tree about which I commanded you, 'You must not eat of it,' cursed is the ground because of you; through painful toil you will eat of it all the days of your life. It will produce thorns and thistles for you, and you will eat the plants of the field" (Gen. 3:17; NIV). Numerous references in both the Old and New Testaments cast thorns in a good light only when they are used in a hedge to confine animals or protect vineyards. Most of the references indicate that people resent having to share the world with them. "For ground that drinks the rain which often falls on it and brings forth vegetation useful to those for whose sake it is also tilled, receives a blessing from God; but if it yields thorns and thistles, it is worthless and close to being cursed, and it ends up being burned" (Heb. 6:7; NASB). The most symbolically potent of these references is the crown of thorns that was placed on the head of Jesus by Roman soldiers. (I'll examine this in the next chapter.)

Of course, thorns were not invented to bedevil humankind. Plants began to evolve them 420 to 370 million years ago, long before our ancestors made their first appearance. One of these early thorned species was *Drepanophycus spinaeformis.* Less than three feet tall and reproduced by spores, it was studded with short, upturned spikes that may have been sharp. These are classified as spines rather than true thorns because they developed from leaves, whereas thorns developed later from branches.[19] Prickles, such as those on the stems of roses, are extensions of the plant's skin. The original purpose of these early spikes is unknown. Scientists speculate that they were devices for collecting dew. Because the swamps where *D. spinaeformis* flourished also teemed with eight-legged crustaceans the length of a horse, the spikes may have helped prevent the plant from being eaten. This is also the prevailing theory about modern spiny plants. However, the precise function and outcome of floral spinescence—the quality of possessing bristles with sharp, stiff ends—are still subject to scientific debate.

Even plants with thorns are eaten. In the spring, deer chew tenderly on our *C. douglasii,* and animals confined by the laid hedges of Britain and Ireland snack on the hawthorns and blackthorns in these living fences. There are pastures in central New York State where dairy cattle have carved topiaries from hawthorns and wild apple trees, which, unlike the cultivars in orchards, have thorns. The distinctively domesticated forms of bowls and cones in these trees are the result of cows "pruning" them by nibbling on the tender tips of the branches year after year. This causes the plants to sprout new growth composed of short, interwoven branches studded with thorns, a dense shell the trees have developed to defend themselves. Constant pruning keeps a tree at the shrub stage for years and has led to the nickname "bovine bonsai." But once the tree eventually rises to a height beyond the reach of the cows, the crown explodes with vertical abandon and an hourglass shape is produced. Hawthorns and apples are usually the only trees that grow in cattle pastures because the beasts eat all the other varieties. However, other woody plants such as chokecherry and white ash will exploit the safe haven these trees provide, and take root inside their shells. After the cattle eat the apples and haws, the seeds are defecated and sprout in the manure. Cattle typically avoid plants growing in or around their own feces, at least for a while, which gives the hawthorns and apples an additional advantage. Cattle may also avoid hawthorn saplings that have not yet grown thorns (our *C. douglasii* sprout thorns when the saplings are about two feet high) because they don't like the taste. A study of browsing habits in Scotland revealed that among the saplings of eleven different trees, cattle rated hawthorns their least preferred—oak was their favorite.[20]

Since a plant invests a lot of energy producing thorns that could otherwise go into fruit and seed production, thus increasing its chances for producing a new generation, Colin Beale at the University of York wondered, "Why grow thorns if they don't work?" Studying the eating behavior of the many kinds of animals browsing the trees of the African savannah, where almost every plant has thorns, Beale noticed that these vegetarians employ two strategies in herbivory (the act of eating plants). They either prune the tree by biting off branches—wood, thorns, leaves, and all—or they avoid the thorns altogether by carefully nibbling the leaves in between them. Giraffes, for example, use their dexterous, twenty-inch tongues to coax leaves from profoundly spinescent acacia trees, which are their favorite food. But looking closer, Beale saw that when an animal was "pruning" it was actually dining only on the ends of the branches where the thorns are still soft.[21]

Beale told me that although spinescence does not prevent herbivory, it inhibits it. "The presence of thorns will slow down intake rates for deer" he said, "and there's good evidence that it does the same for goats and sheep." He pointed out that when researchers removed the thorns from a tree the animals ate more of it, and more quickly. Experiments have shown that a tree responds to a high rate of herbivory by growing more thorns. When fertilizer was given to a tree on the assumption that it would use the extra energy to grow foliage, it grew more thorns. That makes sense if you consider that the more delectable things a tree offers the more likely that it will be eaten. Plants that grow in deserts and savannahs tend to be thorny because it takes so much longer in dry climates for them to regenerate foliage. So what explains the enormous numbers of hawthorns growing across the wet and mild landscapes of Ireland, Normandy, and Britain? These are products of human planting. And they have actually *benefited* from centuries of their association with grazing animals, which have eaten their competition and pruned the hawthorns to make them bushier.[22]

As Beale points out, as long as a plant is going to the trouble and expense of producing thorns it makes sense to warn the herbivores about its arsenal before they discover it by taking a bite. Some hawthorn species do this by producing thorns that are more brightly colored than the branches. The *C. succulenta* I collected in central Montana, for example, not only has a dozen two-inch thorns per foot of branch, the thorns have thornlets, both of which are colored a luxurious shade of burgundy. And they are also varnished smooth to enter flesh with the least amount of resistance. On our *C. douglasii* there are six to seven varnished purple thorns, one and a half inches long, per foot of branch. And on the comparatively delicate *C. monogyna* I looked at in the countryside of southeast Ireland I counted ten curved half-inch thorns per foot, colored reddish-brown at the tip. Since the twigs of our *C. douglasii* saplings are a rich, dark red it might be that deer and our horses confuse the twigs with thorns. This kind of visual warning is called aposematism. A good example of the phenomenon in animals is the black-and-white stripes of the skunk, and the bright bands of red and yellow on the venomous coral snake.

Some species of *Crataegus* take their defensive zeal a step farther

A branch from the fleshy hawthorn, *Crataegus succulenta*. Collected by the author in central Montana, it has retained a few haws and some leaves. The thorns are two to three inches long with additional thornlets. (illustration by the author)

and go on the offense. Researchers in Israel examined thorns collected on Mount Carmel from the spiny hawthorn, *C. aronia,* which produces yellow haws an inch in diameter. What they found alarmed them. The thorns, which are red when young and turn gray when mature, were covered with a biofilm composed of colonies of bacteria attached to the surface and protected by a hard sugar coating excreted by the thorn. Genetic analysis revealed twenty-two species of microbe in the biofilm on the hawthorn and another spiky tree, the date palm.[23] Thirteen kinds of these bacteria colonies are extremely pathogenic to humans and animals, including anthrax, dysentery, tetanus, and gangrene. One might conclude that these trees are armed not simply to deter attackers but to kill them. Adding insult to possibly mortal injury, some thorns harbor fungi that can only infect humans and animals if they get inside the body through a puncture wound. One fungus causes Fonseca's disease, a skin infection found mostly in the tropics that can cause elephantiasis. Another skin ailment common all over the world is rose gardener's disease, which is introduced by a wound from a thorn or a prickle and produces pink and purple lesions.

The thorns of *Crataegus* can also cause synovitis, a form of arthritis that results in the inflammation of the lining of a joint. It is caused when a knuckle, for example, is punctured by a thorn that leaves behind a small bit of itself when it is removed. The tissues react in the same way they do whenever they are invaded by foreign matter—with pain, swelling, redness, and heat. These symptoms might show up days or weeks after the injury. If the thorny material is not surgically removed, the result can be chronic arthritis, the loss of motion, and the eventual disintegration of the joint.[24] This happened to Clara, one of our Border collies, who died of a stroke before anyone could figure out that a spike from one of our hawthorns had pierced her wrist. From then on she had favored that foot so much her shoulder began to atrophy.

As these defensive and offensive strategies were (and still are) devised by spiky plants to protect themselves, herbivores were developing ways to eat them, despite the thorns, because they had to. So the tree grew more thorns or bigger thorns. This strategy is also part of plant-on-plant warfare because, for example, trees in the savannah with less protection will probably lose a higher percentage of themselves to herbivory than those that are more heavily defended. In 1973 the biologist Leigh van Valen likened this escalation of hostilities to an evolutionary arms race. Just as the United States and the Soviet Union engaged in a frantic contest during the Cold War to stockpile the biggest nuclear arsenal, with no benefit to either nation (although inspecting missile silos was the reason my father brought home a government paycheck every week), thorn trees reacted to being eaten by growing daggers and small leaves that were harder to reach. Herbivores reacted by evolving long, thin tongues and specialized foraging behaviors, the trees reacted by dipping their daggers in poison, and so on and on. What became known as the Red Queen Hypothesis proposed that evolution is the result of an organism constantly adapting to its ever-changing environment and to other ever-changing organisms that it is pitted against. All of this just to survive. (The hypothesis name derives from the Red Queen's explanation to Alice in Lewis Carroll's *Through the Looking Glass:* "Now, *here,* you see, it takes all the running you can do, to keep in the same place.") An extreme example is the whistling thorn tree, *Acacia drepanolobium.* Some acacias whose thorny defenses have been overcome or ignored flood their leaves with toxin to dispel herbivores, but this species employs a different tactic. It offers ants sugar-packed nectar and a place to live inside hollow galls that sprout white thorns. In exchange, the ants swarm from openings they have carved in the galls to attack foraging elephants, stinging them on the eyes and the linings of their trunks. The elephants are so terrified of these tiny insects that their mere presence on the tree will frighten the gargantuans away. When researchers in Kenya drove off the ants with smoke the elephants ignored the thorns and ate the acacias with happy abandon.[25]

Another theory holds that the reason hawthorns are eaten these days is because their thorns were evolved to fend off animals that no longer exist. Around thirteen thousand years ago North America was ruled by enormous mammals, including the elephant-like Gomphothere, the wooly mammoth, a three-ton ground sloth, and a twenty-six-hundred-pound camel, in all some forty species of megafauna that are no longer with us. Unlike the white-tailed Bambis that nibble on our *C. douglasii,* delicately avoiding what for them are wide-spaced thorns, these behemoths had very big mouths, indeed. Scientists believe that since they never could have foraged *between* the spikes, and would have been injured if they had tried to gobble an entire branch at one mouthful, they avoided thorny trees like hawthorns to feed on easier fare. Because thirteen thousand years is the blink of an eye in evolutionary terms, hawthorns are still defending themselves against the megafauna. (Maybe this is why they look haunted—they're terrorized by ghosts.)

The same with our old friend, the Osage orange. You may have seen one of these peculiar trees in the fall surrounded by the fruit it's dropped, the grapefruit-sized "hedge-apples," and wondered why animals and birds were letting that food go to waste. The fruit stays where it falls because the only beasts that ever liked to eat it are extinct (in fairness, the seeds are edible if you are willing to spend all day extracting them from the fruit's chemical-tasting pulp). This may be why the natural range of *Maclura pomifera* has been reduced from most of North America to a narrow strip in the South, and the number of *Maclura* species has dropped from seven to one. Another example of this sort of evolutionary anachronism is the pronghorn antelope. I've seen them bounding across the Montana prairie at speeds of fifty miles an hour. Unless the pronghorn just thinks doing this is fun, why would it need to run so fast when the beasts that want to eat it are in comparison foot-dragging wastrels such as wolves, bears and cougars? Maybe it's because the pronghorn is a survivor from an era when it was hunted by *Miracinonyx,* the extinct American cheetah.

Seventy million years ago a mass of land formed by the relentless process of continental drift connected what is now Siberia with Alaska. This area, now called Beringia, meets all the requirements to qualify as a subcontinent. Because the snowfall over Beringia was relatively light, the heart of this cold, arid land was never covered with glaciers during the planet's five great ice ages (technically, we are still living in the most recent of these because of the ice sheets over Greenland and Antarctica). So it became a sort of Shangri-La surrounded by ice, a grassy refuge stretching a thousand miles from north to south, and from Kamchatka to the Yukon, teeming with life. During the coldest periods of these glaciations so much water was tied up in ice sheets the level of the seas was as much as 360 feet lower than it is today. When an ice age ended, Beringia was flooded, but it appeared again when the planet cooled. Similar land masses appeared and disappeared in Southeast Asia, and from continental Europe to Ireland and Britain. Meanwhile, the flora and fauna that had migrated to Beringia from Eurasia made their way into North America. And North American species went to Eurasia following an extended layover in Beringia. The horse and the camel were noteworthy travelers to the Orient, where they flourished while their descendants in North America became extinct. And humans were among the most successful migrants to the Americas.[26]

One of these plants was the hawthorn. During the Eocene epoch, which lasted from fifty-six million to thirty-four million years ago, it is thought that *Crataegus* gradually spread from China to the Americas through Beringia. Again in the Miocene, from twenty-three million to five million years ago, three more waves of migration occurred, and possibly two waves going in the opposite direction. As they do now, hawthorns used birds and animals to get around, coevolving with them in the never-ending dance of the Red Queen. A beautiful fossilized hawthorn leaf dating from fifty million years ago was discovered in a bed of shale in north central Washington State, in a rich fossil site, full of plants and fish, called Stonerose. It interests me to see that it lies on a direct path between Dark Acres and the most northern extent of *C. douglasii,* in the Anchorage area just south of Beringia. In a more scholarly approach to the history of hawthorns in the Pacific Northwest, two things have been suggested. First, the relatively short stout thorns of *C. douglasii* bear a striking resemblance to those of a group of *Crataegus* species from Asia. And second, *C. douglasii* has demonstrated its enthusiasm for travel by making an appearance around the northern Great Lakes, in what are called disjunct populations.[27]

Twenty thousand years ago a small, scattered population of Eurasians is believed to have outlasted the heavy glaciation of the northern hemisphere by wintering in relatively balmy Beringia. They had been isolated from their ancestors for at least five thousand years and presumably had developed very different cultures. This was a period when browsers such as bison, yaks, and mammoths—omnivorous herbivores—transformed Beringia into a sea of grassland in much the same way that the ecology of the Great Plains was developed by and coevolved with the bison. The herbivores were hunted by wolves, bears, lions, and people. When the glaciers retreated enough to open a gateway to the continent, the people followed the animals they hunted as they migrated south and then east. They eventually made their way into Central and South America. Meanwhile, without the herds to manage the grass, Beringia reverted to tundra dominated by mosses and lichens. At least this is the theory about the peopling of the Americas that has held sway for three generations. Another recent theory posits that Beringians in boats traveled down the west coast of North America, subsisting on kelp, seaweed, and shellfish. Yet another idea holds that Eurasians crossed the ice shelves over the North Atlantic and made their way into what is now Canada before spreading west and south.[28]

By whatever means humans made their way to North America, they found many of the same genera of plants that their ancestors had exploited in Eurasia. One of these was the hawthorn. While it was used for food and pharmacy, just as it was on the other side of Beringia, like most living things it became intertwined with the Unseen World, as well, though not to the extent that it did in Europe, where, as we'll see, it became a central symbol in the iconography of Christianity.

E L E V E N

The Crown of Thorns

Miracles are not contrary to nature, but only contrary to what we know about nature.

—AUGUSTINE

On a freezing December night in 2010 a person wielding a chainsaw cut off the branches of a lonely little hawthorn growing on a hill in England, leaving a butchered trunk.[1] While such events are commonplace in the neighborhoods around Dark Acres, where people scavenging for firewood routinely assault the forests, in Britain and other parts of the Christian world this act of attempted arborcide made the front pages. For the victim was no ordinary tree. It was the Glastonbury Thorn, and its story unites Christianity with a much older tradition.

The Gospels of Matthew, Mark, Luke, and John offer accounts of a prosperous and heroic Jewish man named Joseph from the Judean town of Arimathea. After Jesus died on the cross, Joseph went boldly to Pontius Pilate, the Roman governor of the province and reluctant judge at Christ's trial, and received permission to take the body. With the help of another disciple, Nicodemus, Joseph removed the spikes from Jesus' flesh, anointed the corpse with myrrh and aloe, and wrapped it in linen. Then he took the body to a tomb carved out of the face of sheer rock that he had prepared to serve as his own tomb, placed it on a stone shelf inside the tomb, and blocked the opening with a large rock. Joseph of Arimathea was never mentioned again in the New Testament.

But in the ninth century he was back in the news. In *The Life of Mary Magdalene,* the archbishop of Mainz wrote that after these momentous events Joseph boarded a Phoenician ship and sailed from Judea to the mouth of the Rhône in what is now the south of France, and from there around the Hibernian peninsula. His ship entered the Bristol Channel in southwest England and set anchor off the coast of what was then an island. Later accounts of Joseph's introduction of Christianity to the Celtic tribes of Britain claim that he was in possession of the Holy Grail, the chalice used at the Last Supper and the vessel it was believed Joseph used to collect the blood of Christ after he had taken the body from the cross. Once Joseph was ashore he climbed what is now Wearyall Hill. Exhausted, he plunged his staff into the ground and lay down to sleep. When he awoke, the staff—apparently crafted from a Mediterranean species of hawthorn—had burst into full white bloom. In 673 a Benedictine monastery was founded at Glastonbury on the site of the modest church purportedly built by Joseph, and the walls of this abbey were built around the hawthorn, which was unique among species of *Crataegus* because it bloomed twice a year, once in the spring and once at Christmas.[2]

As the centuries passed, Joseph's stature grew. In 1191 the monks announced that they had exhumed a tomb containing two bodies that had been placed in a hollowed-out log and buried fifteen feet deep between two small stone pyramids in the abbey's cemetery. A leaden cross interred with them identified the deceased: "Here on the Isle of Avalon lies buried the renowned King Arthur, with Guinevere, his second wife." A tress of blond hair was also found in the coffin, but it disintegrated when one of the monks reached out to touch it. The tomb's discovery lent credence to the belief that the island Joseph had landed on was indeed Avalon, the place where Arthur was taken after being fatally wounded during the battle against his nephew Mordred, the traitor. Arthur was claimed to be a direct descendent of Joseph, who was himself either the Virgin Mary's uncle, or Joseph's uncle. The king and the apostle were further linked because the man from Arimathea was the keeper of the Grail, the holy object sought in acts of chivalrous piety by Arthur's knights.[3]

Such legends have strong narrative force. So powerful is Joseph's story, in fact, that an epic film, *Glastonbury: Isle of Light,* is scheduled for international release in 2015. A moody, evocative graphic for the movie shows the Glastonbury Thorn on Wearyall Hill (before the tree was vandalized, of course) set against a horizon where the tower of a ruined medieval church, built on a high tor, looms above the village. The plot revolves around Joseph's voyage from Judea with the Grail, his landing in Britain, and his alliance with the Celtic warrior Caratacus to battle the Roman invasion ordered by Emperor Claudius in 43. At one point Liam Neeson was being considered for the lead.[4]

Outside of the Gospels, which may have been collections of Jewish anecdotes rather than an account of real events, the entire life of Joseph of Arimathea was confected by the monks of Glastonbury Abbey and writers who had little regard for the distinction between fact and fancy. When the Benedictines made their announcement about finding King Arthur's tomb they may have been following a typical practice at the time in which monks concocted legends about the founders of their monasteries in order to attract pilgrims and the patronage of the nobility. The abbey had been badly damaged by fire in 1184, and though Henry II financed its reconstruction, that source of revenue dried up when he died in 1189. His successor, Richard I, was interested in the abbey only as a source of funds for his Crusades. While the British eagerly accepted the story about Arthur because he was deeply ingrained in Celtic and later Saxon lore, when the Benedictines tried fifty years later to link Joseph of Arimathea to the abbey's founding, citing the legends of the Grail that named him the Apostle of Britain, their monastic propaganda proved to be a much tougher sell. For one thing, none of the early historians of Christianity in the first centuries after Christ had mentioned Joseph's association with the Grail or his introduction of the church to Celtic Britain. In fact, it wasn't until the fifteenth century that he was added to the saints tied to Glastonbury and finally achieved some fame. His connection to Arthur, whose tomb has never been verified by anyone other than the monks, was a confection penned by Robert de Boron, whose twelfth century *Joseph d'Arimathie* described Joseph's voyage from Jerusalem with the Grail. Nonetheless, the abbey prospered again and became a major landowner in the county of Somerset. But in 1536, Henry VIII dissolved Britain's eight hundred monasteries and confiscated their land and riches, leaving fifteen thousand monks and nuns homeless. The abbot of Glastonbury, Richard Whiting, was declared a traitor for resisting. On a bleak November morning in 1539 he was tied to a wooden travois and dragged by a horse through the streets of Glastonbury to Saint Michael's tower on the summit of Glastonbury Tor. There he was hanged. Then, in the typical overstatement that characterized punishment in the Renaissance, he was beheaded and his body cut into four pieces. His head was impaled over the gateway of the abbey as a warning to other seditionists.[5]

As for the Thorn, botanists are of two minds. It might have originated in the Mediterranean, whose climate encourages hawthorns to flower twice a year. However, a twice-blooming cultivar of this common English hawthorn, *C. monogyna* 'Biflora,' was found growing wild in a nature reserve a hundred miles north of Wearyall Hill. Whatever its origin, references to the unusual hawthorns began showing up regularly in the sixteenth century. For example, "there is an hawthorne which is grene all the wynter," William Turner wrote in his 1562 *New Herball.* No records have been found showing how long this Biflora cultivar has been flourishing at Glastonbury, but the first reference to its origin as Joseph's walking stick did not appear until the eighteenth century. The date of its appropriation and promotion by the abbey is also unknown. Regardless, the Glastonbury Thorn is one of Christianity's most compelling icons in Britain, and believers take its story as an article of faith.[6]

For many, therefore, the assault on Wearyall Hill was a heart-wrenching sacrilege. The day before the little tree was vandalized, an annual ceremony had taken place in which a schoolboy climbed a ladder and cut a budding bough from a Biflora growing in the yard of the fifteenth-century Church of Saint John that would be presented to Queen Elizabeth to decorate the royal table on Christmas Day (there are several such cultivars growing in and around Glastonbury). This tradition is a century old, but bough-giving dates from the reign of James I, five centuries earlier. "The vandals have struck at the heart of Christianity," said Katherine Gorbing, director of Glastonbury Abbey. "Like the whole town, we are shocked and appalled." And this was not the first time the Thorn had been attacked. It was chopped down and burned during the English Civil War in the 1640s by Oliver Cromwell's Roundhead soldiers because they believed it incited demonic belief in magic and superstition. But the residents of Glastonbury gathered roots and cuttings from the tree and used them to grow new thorns in secret places. (If they used cuttings they would have had to graft them onto the root stock of ordinary *C. monogyna* because hawthorn grown from the seeds or shoots of Biflora bloom only once a year.) As for the Biflora on Wearyall Hill, the spot Joseph was reputed to have visited, the tree was hardly ancient. It was planted in 1952 to replace its predecessor, which itself had been planted the year before to commemorate the Festival of Britain. The first thorn did not survive a drought.[7]

No one knows the motives for the assault on the thorn, since the culprit as of this writing has not been caught, but speculation has ascribed the vandalism to militant atheists, militant pagans, or mindless vandals. (Another theory is that it was someone trying to punish the owner of Wearyall Hill, the major shareholder in a financial company whose collapse a couple of months previously had led to his arrest the week of the attack.) The Wearyall thorn was bandaged, and prayers were said for its restoration. Like most severely coppiced trees, all it really needed was water and sunlight to begin growing with even more force than before it was savaged. But misguided visitors poured honey and Guinness on its roots, tied ribbons to its trunk, blocking the sunlight, and drove coins into its bark for good luck. Even then, it managed to grow a few new sprouts. But trophy hunters stole them, along with strips of its bark. The beleaguered hawthorn was replaced in 2012 with a new Biflora grafted at Kew Gardens in London and planted to mark the Diamond Jubilee of the queen. There was talk of installing one of England's ubiquitous security cameras to watch over the sapling. Two weeks later the stem was snapped off eighteen inches from the ground. Yet another tree was planted, this time in the heart of the village, where it was believed vandals would be deterred. But again, in June 2013, the trunk was partly severed and, again, snapped off.[8]

The Biflora growing in the churchyard of Saint John was planted during the 1930s by George Chislett, the abbey's head gardener. His son took scions from this tree, grafted them on the rootstock of blackthorn (*Prunus spinosa),* a rose family cousin, and sent them to arborists all over the world. A Biflora from Glastonbury is flourishing in the garden of the National Cathedral in Washington D.C., which supplied a California nursery with scions that are grafted onto *C. monogyna* rootstock and were being sold for thirty dollars apiece until increased demand compelled the company to refuse new orders. And the cathedral sent scions to the town of Glastonbury, Connecticut, so that the community could have a living source for the image of the hawthorn branch bearing fruit that adorns its municipal seal. Despite assaults on it, the tree that must have filled viewers with wonder as it blossomed in falling snow is not just surviving, it is multiplying. Meanwhile, the same cannot be said for the faith that gave the tree new life generation after generation. The dramatic declines in membership in both the Catholic Church in Britain and the Church of England have led some observers to wonder whether Christianity is dying out in the British Isles.[9]

While the pagan spirituality that Christianity supplanted has largely faded, one ancient Celtic observance is undergoing something of a revival. Beltane was a raucous gathering at around the first of May that marked the beginning of warm weather, when the cattle were driven to their summer pastures. During this celebration of the rebirth of nature, huge bonfires were lit as part of the notion that smoke and ashes had protective powers, and people danced and led their cows between the flames. In Ireland, celebrants adorned a hawthorn called a May Bush with shells and ribbons, scattered hawthorn blossoms on their doorsteps for luck, and visited holy wells and "rag trees," wishing trees to which supplicants tied bits of cloth and other "presents." The revival began in the 1980s when hippies, neopagans, and Wiccans began to build bonfires to celebrate the return of summer in places such as Edinburgh and Glastonbury, where they greet May Day on the Tor with drums and chants and dancing. So-called Celtic reconstructionists try to observe Beltane with as much historical accuracy as possible, celebrating the holiday only when the hawthorns are in bloom.[10]

While the blossoming hawthorn is a symbol of fertility, in the Balkans the tree was an emblem for the cult of the undead. During the early 1880s in the Slovakian village of Tomiŝelj, for example, villagers reported strange visits from a prominent man who had died. His wife claimed that he stole into her bed at night, and his neighbors said they often saw him sitting on a rock. The village priest and a parishioner dug up the man's grave and determined that he was a vampire. They drove a stake made of hawthorn through his heart, and re-covered his grave. In

1882 vampire hunters in Varna, Bulgaria, opened the grave of an undead person suspected of causing an outbreak of disease and discovered that the corpse was full of fresh blood. They thrust a thorn from the tree into its breast. Then, for good measure, they cremated the body on a pyre of hawthorn branches. Apparently, a special kind of hawthorn is needed to impale the undead. According to an "instructional manual," *South Slavic Countermeasures Against Vampires,* "The hawthorn must have been grown in the high mountains in a place from which the bush could not have seen the sea." Because it was thought that vampires were allergic to the thorns of hawthorns owing to its association with the crown of thorns, in eastern Serbia a small hawthorn peg driven into the grave next to a cross was thought to prevent the corpse from turning into a vampire, while tossing hawthorns on a grave would prevent any dead person from taking revenge on the living.[11]

Before Christ was crucified he was mocked. According to the Gospel of Matthew, Roman soldiers took him into the headquarters of Pontius Pilate, stripped him, dressed him in a scarlet robe, and put a reed in his right hand. They twisted thorn branches into the shape of a crown, and put it on his head. Then they knelt before him and cried, "Hail, King of the Jews!" (According to the Gospel of Mark, the robe was purple.) The scene is rich in symbolism. God punished Adam for eating the apple by turning him into a farmer whose grain would have to compete with thorny plants. And Roman citizens were awarded what was called a civil crown made of oak leaves as recognition for their bravery in battle. Although none of the Gospels mentions what kind of thorns were used, Europeans filled in the blank with hawthorns. From one end of the continent to the other the tree had been associated with the Unseen Force long before Christ arrived on the scene.

An embellishment of the New Testament account of the crown of thorns appeared in a wildly popular book published in the fourteenth century. Written in Anglo-Norman French and translated into a number of languages, *The Travels of Sir John Mandeville* was a wholly unreliable and fanciful account of a journey to the Holy Land, Egypt, India, and China that may have been the work of a Benedictine monk named Jan de Langhe, who was an avid collector of travelogues. It contains accounts of dog-headed men, cannibals, and Amazons. In his second chapter, "Of the Crosse and the Croune of oure Lord Jesu Crist," the author discounts the notion that the crown was made of thorns, claiming that he had seen it with his own eyes and it was made of the spines of what is now called the sharp rush, *Juncus acutus,* a salt-tolerant plant growing on the shores of Israel, among other places, whose leaves narrow to a flesh-piercing point. The author claims to have owned a spike from the crown, which resembled the spike of a whitethorn, or *C. monogyna,* perhaps to explain to his readers why many had assumed that the crown was made of hawthorn.[12]

The author also writes that before his arrest, Jesus was led into a garden where an "albespine" was growing. This is one of the English names for *C. monogyna.* In the garden was a group of Jews who scorned Jesus for claiming he was their king and who crafted a hawthorn crown, which they clamped on his head, drawing blood. Then Jesus was led to another garden, where authorities crowned him again, this time with "sweet thorn," probably the sweet thorn, *Acacia karroo,* which is native to southern Africa. The author obviously liked his idea so much that he added visits to two more gardens. In the first Jesus was crowned with eglantine, *Rosa rubiginosa,* a western Asian rose bristling with sharp prickles. Finally, he was fitted with the crown of thorns made of the sharp rush that he wore on the cross. How readers reconciled these various crowns with the single crown mentioned in the Gospels is unknown. When citing this ludicrous account many sources omit the other species and mention only the whitethorn, as in, for example, an 1889 weekly periodical, *All the Year Round,* founded by Charles Dickens, and edited by his son after his father's death in 1870. An article in the journal notes that the use of hawthorn in the crown explains why the French believe that on Good Friday hawthorns utter cries and groans.[13]

Azarole (*C. azarolus*) is a species of hawthorn native to Israel that could have been used by a Roman soldier willing to go to the considerable trouble and probable pain of crafting a spiky crown. Another native species is the Sinai hawthorn, *C. sinaica,* a hybrid of *C. azarolus* and *C. monogyna,* which also grows in the Levant. *C. sinaica* is so rare these days that scientists are experimenting to see whether they can grow masses of seedlings in test tubes.[14] Circumstantial evidence indicates, however, that the crown of thorns, if it ever existed, was not made from any of the plants I've mentioned so far.

Aside from the Gospel accounts, the crown was not heard of until the fifth century, when it was listed as part of a collection of Crucifixion relics housed in a Jerusalem church on Mount Zion. Around 570 a Roman statesman and writer named Cassiodorus claimed to have seen the crown, "which was only set upon the head of Our Redeemer in order that all the thorns of the world might be gathered together and broken." Undermining the authenticity of this relic, Bishop Gregory of Tours wrote in the sixth century that the thorns were green and miraculously turned fresh again every day (there is no evidence that Gregory ever left France). In 870 a monk wrote of seeing the crown on Mount Zion. About 1063 it was moved to Constantinople, where it came into the possession of Baldwin II, ruler of the Latin Empire of Constantinople from 1228 to 1261 during a period when the financially beleaguered empire did not extend much beyond the city limits. At one point Baldwin was reduced to selling the lead sheeting on the roof of the palace. In 1237 the city fell under siege by an alliance of belligerents. While Baldwin was off begging for money in the courts of Europe, his barons pawned the crown of thorns to a consortium of Venetians for 13,134 gold coins. They pawned it again to Venetian banker named Nicolò Quirino on stricter terms. He could keep or sell the relic if they did not repay the entire sum plus 12 percent interest within four months. Baldwin begged the king of France, Louis IX, for help and was rewarded when Quirino agreed to part with his new treasure.[15]

The crown was transported to France in a wooden trunk, and in 1248 was extravagantly displayed at Sainte-Chapelle, considered one of the pinnacles of Rayonnant Gothic architecture, which was built in Paris by Louis IX to house his collection of holy relics. These included the lance thought to have been driven into Christ's flesh, a hunk of wood believed to be part of the cross, and the sponge that was soaked in vinegar and offered as a mockery to the thirsting Christ while he was dying. Now Louis added his jewel, the crown of thorns. There it stayed until the French Revolution, when it was moved to the National Library, and finally to Notre Dame in 1806. In 1896 it was encased in a tube of crystal and silver inlaid with intricate gold filigree. (Such was the common fate of objects associated with Christ the simple carpenter, who preached that it was not possible to serve God and Mammon.) The first Friday of every month and several times during Lent the crown is removed from its ornate gold reliquary and paraded around the chapel. What you can see of it is a braid of plant material said to be rushes or canes, twenty-six inches in circumference, held together with gold thread. It was around this braid that thorns or branches bearing thorns were attached. But this bound and imprisoned relic no longer has any thorns. One by one over the centuries they were removed and given away as favors by its owners. Louis IX and his successors distributed sixty or seventy of them. One of these, presented to Mary, Queen of Scots, has been housed for four centuries at England's Stonyhurst College college inside a small glass-and-metal tube that also contains the beheaded queen's pearls.[16]

Modern examinations of some of these surviving thorns have led to the conclusion that they come from a spiny, twenty-foot tree that spread into Israel from the Sudan called Christ's thorn jujube, *Ziziphus spina-christi.* Staying green all year except in especially chilly winters, it has a pair of aggressive inch-long thorns at the base of each leaf, one straight and one hooked, an arrangement designed to jab a predator coming and going. One piece of jujube thought to be part of the original crown is housed in Italy, and another in Belgium. A study of the Shroud of Turin, believed by the faithful to have been the linen Joseph of Arimathea used to wrap Christ's body, revealed pollen said to be from the jujube, as well as from a spiny tumbleweed, *Gundelia tournefortii.* And the head of the man whose ghostly image is pictured on the fabric shows wounds that could have been made by thorns, possibly caused by two spiky crowns. Unlike the simple circlet suggested by the relic housed at Sainte-Chapelle, the evidence of the wounds revealed on the shroud suggests that the crowns were shaped more like helmets that covered the entire skull. Another avenue of research concluded that the shroud came from the vicinity of Jerusalem, was first used in the months of March and April, and dates to some time before the eighth century. Despite controversial radiocarbon tests that established its origin in the Middle Ages, and pigment analysis suggesting that the reddish stains were paint, not blood, the shroud remains one of the world's most alluring mysteries.[17]

The hawthorn's white flowers, and the fact that it is the first tree to bloom in the spring, led to its becoming associated with the Virgin Mary. The tenth-century mystic Hugh of Saint Victor wrote with the typical anti-Semitism of the era that Mary was the flower blooming out of the thorns of the Jewish race, the hawthorn's blossoms symbolizing her perpetual virginity. And so the hawthorn became an avatar of Marian apparitions, visions of the Virgin that appeared on or near the tree. A legend from northern France, for example, holds that shepherds came upon a burning hawthorn in which was a statue of the Mary holding her Son. The Basilique Notre-Dame de l'Épine (Basilica of Our Lady of the Thorn), a masterpiece of the Flamboyant Gothic style, was built around this statue and completed in 1527. In 1399 in Spain two shepherd boys saw a group of ghostly people gathered around a hawthorn on whose crown was a lady glowing so brilliantly that they were forced to avert their eyes.[18]

In November 1932 five schoolchildren were playing in the yard of a convent in the farming town of Beauraing, Belgium, when a vision appeared to them, floating on a cloud near a railway bridge across from the school. It was a "beautiful lady," the children said, dressed in a long, white gown and a veil, radiating a blue light. No one believed them because the children were the only people who saw the vision, and they were known for pulling pranks. But they were examined by doctors, and questioned by the authorities, who found no evidence that they were joking. The children, who had apparently fallen into a kind of rapture, told identical tales of the vision. Over the next six weeks the apparition of the Virgin appeared to them more than thirty times. Most of these visits occurred under the branches of a hawthorn tree near the entrance to the school garden, and sometimes it was preceded by a flash of fire.[19]

One of the girls who first saw this vision, Andrée Degeimbre, went to the hawthorn to pray over her rosary, rain or snow, every evening for forty-five years.[20] In 1949 the Catholic Church officially recognized the supernatural character of the apparitions, and in 1985 Pope John Paul II knelt before a white marble statue of Mary that had been placed under the miraculous Beauraing hawthorn.

In Ireland a lonely little whitethorn stands in the middle of nowhere bedecked with rags, shreds of clothing, and plastic bags, like a Christmas tree decorated by the homeless. It grows near the eight-hundred-foot summit of the Knockmealdown Mountains in County Waterford, right next to a two-lane highway, and it is what is known as a "rag tree," one of scores of such devotional trees that can be found from one end of Ireland to the other. (In the north of the island these are called "clootie trees.") This particular thorn is on the route of an old fifty-mile pilgrimage from Ardmore on the coast of County Waterford to Cashmel in County Tipperary. Pilgrims stop to add something to the tree on their way to visit nearby Melleray Grotto, where in 1985 visitors witnessed a statue of the Virgin wandering through the trees of the sanctuary, revealing future events, and conveying messages of peace and love. Pilgrims had adorned the tree with financially worthless but personally meaningful objects meant to petition the Unseen Force for the restoration of health, for riches, or for whatever else they desired.[21]

In Scotland such hawthorns and other species are called "wishing trees." One of the more notable of these is an old hawthorn growing on a path above Ardmaddy Castle in Argyll. Hundreds of coins have been pounded into its trunk, and it is considered good luck to add a coin to the tree and bad luck to take one.[22] When this beleaguered thorn finally dies the cause of death will probably be metal poisoning. A few miles from my great-grandfather's parish stands another lone whitethorn. This "mass bush" grows at a crossroads not far from Killinaspick Church in the Walsh Hills of County Kilkenny. Until recently people tied ribbons to it for luck. From the high ground here you can see down to the River Suir, parts of four counties, and the fields defined by hedgerows where my mother's family tried and failed to make a living. Although memories of the meaning of this tree are fading, during the era following the English Civil War and the defeat of Catholic forces at the hands of the Protestants in 1690, the bush was a focal point for the secret masses celebrated by a furtive congregation whose church had been confiscated by Protestants, like most Catholic churches in Ireland.

Although in Ireland it is considered good luck to hang a sprig of hawthorn over an outside doorway to thwart malign spirits, it is very bad luck to bring the flowers inside. There are several reasons for this suspicion, none of them, of course, verifiable. First, their odor tends to be putrid, reeking like the bodies piled in the street during the Black Death, which entered Ireland through ports such as Waterford. This was not an agreeable smell, especially in the tiny rooms of the houses of Irish peasants. Additionally, the Irish did not want to offend the faerie folk, who made their homes in hawthorns. The tree's Catholic connotations also worked against it. Catholics, who made up 80 percent of the population of Ireland when my great-grandfather was growing up,[23] were so intimidated by the Protestant minority, and the soldiers and police who kept it in power, that they were fearful of bringing hawthorn blossoms into their houses to adorn their home altars. Devoted to the Virgin Mary, these altars consisted of an image of the Virgin set on a table piled with white flowers. The family would gather around it and pray the rosary, which included ten "Hail Mary" recitations. This tradition dates back to the Dark Ages, and spread across Europe with the growing cult of the Virgin, who became known as the Queen of the May. The delicate white hawthorn blossoms symbolized purity (although they can smell like sex as well as death), and the thorns are reminders of the crown of thorns and God's punishment of Adam and Eve for sinning.

In every culture and in every era trees have been worshiped because they were believed to harbor spirits who came and went as they pleased, or served as halfway houses for human souls on their way from death to reincarnation. When the Celts took possession of Ireland they inherited an animistic belief in the power of trees from earlier cultures, and put their own spin on it. They gathered in forests and groves to petition some ninety male and female deities, led by druids, who might train for twenty years in this non-literate culture to advise warrior kings and to commit vast epic poems to memory. To undermine the priestly class, the first Christians in Ireland portrayed them as unremarkable sorcerers without much power. But at the same time they co-opted aspects of the pagan faith. Celtic tree worship, for example, was subverted into a new interpretation of trees as symbols of the might of Christ. The priests of the new Celtic church transformed a sacred meeting place in the forest from one with druidic meaning to one that drew its power from the Bible. They appropriated the boastful stories of druidic magic and turned them into miracles resulting from prayer and faith in a Christian god. The Glastonbury Thorn is a powerful example of a Christian miracle manifested in a tree, and its story was probably adapted from a pagan tale. In the County Roscommon town of Oran is a whitethorn called Saint Patrick's Bush growing beside a spring; both are regularly visited by pilgrims. Saint Patrick reportedly stopped there to rest and gave the place his blessing. Archaeologists have determined that this was the site of an ancient Celtic festival called Lughnasa, which celebrated the beginning of the harvest season.[24] (The tradition of tying ribbons to the bush has faded.) In the story of another purloined miracle, Saint Ita (475–570) pulled a thorn from her donkey's foot. Commanding the thorn never to harm her mount again, she thrust it into the ground, where it grew into a tree whose thorns all pointed downward.

On the continent, Christianity took the opposite approach and tried to eliminate pagan nature worship. As early as 380, Christians were busy destroying Celtic and Germanic trees and forests, and they continued the practice well into the eleventh century. From the twelfth to the eighteenth centuries the Catholic church continued to discredit sacred trees by imprisoning them in masonry to prevent people from adoring them, or subverting them by fixing pictures of the holy family to their branches. In Ireland, most sacred trees are *C. monogyna,* and they grow next to wells or springs, which are considered even more potent than the thorn. There are some three thousand such seeps, including one at Mothel, which were the water shrines of the old Celts. These shrines are often way stations from one holy site to another along a route called a "pattern" (from the old Gaelic word *patrun,* or "patron saint").

In every culture in the world legends can be found of a parallel society of beings living among us, invisible unless they want us to see them. This belief flickers on in Ireland, although it has largely died out in the rest of the industrialized world. The Celts believed that their arrival in Ireland drove a race of beings somewhat smaller than humans into hiding. Eager to capitalize on the existence of faeries, the church told the pagans that these wee folk were confused angels who had been expelled with Lucifer after his heavenly revolution because they could not decide which side to take. But as the angels rained down on their way to hell God decided to send them to Ireland instead. They took up residence in lone hawthorns growing in isolation away from the thickets where hawthorns generally flourish. While people avoided saying the word *fae-*

rie, using *them* instead, they also referred to a solitary thorn as a "noble bush" so as not to draw the attention of these troublesome and often malevolent sprites. Part of the respect people have for lone hawthorns is the fact that they have managed to survive drought, flood, fire, plows, grazing animals, and spades without reinforcement from other hawthorns. The groves at Dark Acres, for example, have developed their own little ecosystems that discourage herbivory and suppress ground cover that could catch fire and damage the trees. They have been flooded by rising river water twice in two recent decades, and were probably stressed by the experience, but none of them died. Plowing these groves would be difficult because of the fortress-like root system the hawthorns have put down.

The strong emotional attachment of some Irish to faerie bushes was evident in 1999 when the proposed route of a new road was altered because the original plan called for it to be constructed through a solitary hawthorn. This was the Latoon Thorn in County Clare, growing in a spot where some believed the faeries of the province of Muenster gathered to discuss tactics before marching off to do battle with the faeries of the province of Connacht. A protest campaign was organized by a folklorist named Eddie Lenihan, who warned road authorities about the consequences of harming this famous *sceach* (Gaelic for "whitethorn"), including but not limited to car crashes. In the end the road was routed around the thorn, and a wooden fence was built to protect it. In 2002 a vandal whacked off every one of the Latoon Thorn's branches, but the tree survived.[25]

In *Whitethorn Woods* (2004), the prolific Irish writer Maeve Binchy tells the story of a spring and a statue of Saint Ann in a grove of whitethorns that lie directly in the path of a road proposed by planning authorities. The town's priest would like to see the idolatrous shrine destroyed, but some townspeople are opposed to the scheme because their prayers have been answered, or they felt that someone was listening to their petition as they tied their strips of cloth, notes, and ribbons to the branches of the thorns. The novel is a study of the power these places have over the imaginations of contemporary people and the conflict between the new and urbane in Ireland and the old and rural, a struggle whose outcome was decided three generations ago in America. In the end, a new route for the road is negotiated and the shrine is spared.[26]

In 1978 the American automaker John DeLorean began construction of a plant on the outskirts of Belfast to manufacture his dream car, the stainless steel, gull-wing, two-seat sports car that was featured in the film *Back to the Future.* In 1982, a year after the first of nine thousand so-called DeLoreans were built, the company, in financial ruin, was shut down by the British government, which had funded the scheme in order the bring jobs to impoverished Northern Ireland, at that time suffering an unemployment rate of 20 percent. That same year, DeLorean was arrested in a $24 million cocaine-trafficking case (a charge he later beat in court). The plant closure cost two thousand workers their jobs. In Ireland it is widely believed that if DeLorean had tolerated the presence of a faerie tree growing from a mound on the grounds of the plant, he would not have had all these troubles. None of his workers would remove the whitethorn, which was in the way, and it was rumored that DeLorean himself destroyed it and the mound on which it grew with a bulldozer in the middle of the night. Regardless of who was responsible, dawn's early light revealed that the tree had simply vanished.[27]

While most of the Irish no longer give any credence to the island's old superstitions, seeing them as a relic of an ignorant rural culture, the belief in the otherworld obviously is still alive. Unlike my Irish ancestors I'm not obsessed with legends about the power of hawthorns. But I have taken note that neighbors of ours who destroyed hawthorns on their land have suffered one catastrophe after another. And I believe that a large thorn living at Dark Acres holds some sort of sway over me, demanding if not my fealty, at least my attention.

TWELVE

The Warrior Queen

A forest of these trees is a spectacle too much for one man to see.

—DAVID DOUGLAS

Leave it to the French to coin an adroit phrase for the obscure and fleeting delusion that although you must have been in a particular place many times before, nothing about the experience seems familiar. *Jamais vu,* they call it. *Never seen.*

Fifteen years after that revelatory spring day at Dark Acres when I discovered the world of the hawthorn, I stood in the rain, nipping hawthorn-flavored vodka from a sippy cup when I suddenly felt as though I'd been transported to another planet a million years in the future. Or dispatched by some time warp to the tenth century in Siberia.

But this was clearly our forest, mine and Kitty's, surrounded by our swamps. And these were our little stock dogs, Clara the Border collie and Lyndon Baines Johnson the Corgi, huddled against each other in the smudged snow, studying me, blinking against the foul weather and wondering when I was going to walk them back through the woods to the house for their dinner. They were pouting because they'd been left alone so much lately.

I shook my head, and rubbed my hands together. After a moment, the jamais vu sensation fled. I tried to summon up its texture, but that was gone as well.

Then something real took over. The panic I'd been pushing away all week started pushing back. The day's last light squeezed through the leafless canopy above my head, the gnarled, thorny limbs drooping like the hands of hags in a corny horror movie. I checked my cellphone again, to make sure it was on. To make sure there wasn't a message from the hospital.

When I was in first grade my mother taught me how to pray—the words you said, the supplicant's tone. I developed a growing list of people and animals that every night I asked God and Jesus to watch over. My petition was sealed and delivered heavenward with the sign of the cross my mother showed me how to make. When I executed this secret, potent gesture—which I tried to do the way the priest did it at mass, with a little kiss to the hand at the finish—I could smell the incense, hear the Latin.

The list started with my mother, of course, and ended with the latest wild beast I'd seen on *The Wonderful World of Disney* on Sunday evenings. It was only after I recited my pleas for these others that I allowed myself to ask for something for myself.

One day I lost a pencil my grandfather had given me. That night, following my request for the well-being of tigers, I asked if I could have a little help in the matter of retrieving this prize. As I was walking to school the next morning my prayer was answered. The fact that the pencil had dropped from my book bag into the street and had been snapped in two by a car rolling over it didn't shake my faith. It just meant that my definition of the thing I wanted had been incomplete. The picture I'd made of it was flawed.

A month later my mother suddenly died. An accident, they told me. A World War II officer who had served with the U.S. Army in the Philippines, she drove every weekday morning before dawn across town to the Air Force base where she worked as a registered nurse at the base hospital. When it was cold she started her old car in the garage attached to our tiny house, let it warm up while she opened the garage door, and went back to the kitchen for a second cup of coffee. But on this January day, after she started the car she somehow slipped getting out and hit her head hard enough to knock her so silly she got back in the car and passed out. A babysitter hired to take care of my kid sister while I spent the day in first grade discovered my sister and me a couple of hours later, in bed, still asleep, groggy and confused. The woman apparently did not smell the exhaust fumes or hear the car running in the garage.

Years later I heard another story. Depressed by her recent divorce from my father, my mother had simply walked out the backdoor of the house, opened the backdoor of the garage, shut it, got in the car, and turned the key.

While I was in college my father told me a third story, a variant of the first. He said the thing she'd tripped over when she hit her head was my bicycle.

For a while after she died I figured she'd return. Maybe not exactly as she'd been. Maybe even in another form entirely. Who knows, I thought, why not a tiger? But when I finally gave up believing she was going to come back, I decided that my apparent failure to imagine her correctly in my prayers—to visualize a whole and vibrant Momma, instead of a two-dimensional sketch of her, a television image—had not caused her death. Nor was it preventing her resurrection. So I cut my losses and stopped praying.

Now, standing at Dark Acres in the tangle of dripping hawthorns, I was thinking about prayer again, for the first time in fifty years. I was thinking about an old joke: a kid with a paralyzed arm lifts his closed eyes to heaven and whispers, "Please, God, make my arm like my other arm." And God does. Suddenly *both* the kid's arms are paralyzed.

A few years ago I came across a paperback in a used bookstore. It was about the landmarks that some Northern Plains Indians believe generate a special power. I'd never heard of these places, even though they all lay within a day or two from the house where I grew up, and from Dark Acres, as well. I wondered whether I might feel something in these sacred places.

And I did, although I still haven't been able to figure out whether the electricity I sensed pouring from those ancient sites was hallucinatory, self-delusional, or a product of the hypertension that runs in my family. Nonetheless, I brought back something physical from each of them—stones from Bear Butte, Dinwoody Canyon, Chief Mountain, and the Sweetgrass Hills; a broken arrowhead from the Ulm Pishkin; a deer bone from the slope below the Medicine Wheel in northern Wyoming. Later I mortared this rubble into our garden wall. The California poppies I planted above this wall flood the stone with a cascade of tiny yellow blossoms.

Now I had a rag tree.

Tied to the branches of the tree were ribbons, strips of paper, shreds of cloth, and snapshots in baggies, which some stranger stumbling into the grove might interpret as offerings, or fetishes. I have to admit that they *are* sort of creepy, like props from *The Ring,* say, or *The Blair Witch Project.*

But they aren't bribes. I mean, the tree has no use for this stuff. While I understand that hanging junk in the woods is as irrational as buying lottery tickets or forwarding chain letters, until events prove me wrong I'll regard the practice as a form of hedging my bets. After all, this hawthorn is considerably bigger than I am. And she is much older. I know this because although I would never dare cut down one of her seven trunks to count her rings, or use a boring tool for the job because boring sometimes introduces diseases, I did use another method to estimate her age. I measured the circumference of the tree's largest trunk at fifty-four inches above the ground. I sawed apart a number of dead branches in order to count the average width of the rings inside. Then I used a formula to figure out the diameter and radius of the trunk, subtracting a quarter-inch for the bark, and divided the radius by the width of an average ring. The numbers suggested that this tree is 120 to 160 years old.

And though I'm a man of no spectacular achievement who shuns his neighbors, belongs to no civic organizations, and only exits the gates of his property to play tennis or shop for groceries, this hawthorn is a member of a feared, venerated species that played a quiet but significant role for thousands of years in cultures all over the world. Despite the fact that hawthorns even figure heavily in the symbolism of my abandoned Catholicism, I feel that connecting myself to them might have value; at least they're living, tangible beings with verifiable powers, unlike those guy-in-the-sky deities.

And although things weren't going well at the moment—in fact, our situation couldn't get much worse—I chose to believe that this old hawthorn had been responsible for some minor good news we received a month earlier, not long after we tied a green sash to its branches.

Green for money. Although quickly spent.

I reached into the pocket of my slicker and withdrew a necklace.

It was a thin silver snake chain bearing a pendant cast into the stylized image of a Hopi quail, inlaid with turquoise and jade. I'd bought it for Kitty in a shop on the Plaza at Santa Fe. She called it her brave bird and wore it when she competed in rodeos as a barrel racer to remind herself to ride hard and cowgirl up.

Avoiding thorns, I reached above my head, draped the brave bird around a branch and shut the clasp. I didn't know what else to do. There was nothing else I could think of to help her.

My cellphone suddenly rang, its vapid little sing-song bringing the dogs to all-fours. I fished it out of my pocket, hoping it was a wrong number. Or a call from a telemarketer. But it was neither.

I named my formidable hawthorn Maeve (pronounced "Mah-ay-vah"), after an Irish warrior queen, to distinguish her from the 135 other hawthorns at Dark Acres. Queen Maeve may not have existed, or the personage we know of may be an amalgam of more than one historical character. Central to the Ulster Cycle, a body of Irish legends and heroic mythology, is "The Cattle Raid of Cooley," which describes how Queen Maeve (or Medb, in Old Irish) led an army from what is now County Connacht against the forces of Ulster to steal a prized bull. (In the decentralized, pagan society of the island during the pre-Christian Iron Age, when some scholars think these sagas were first told, cattle were often used as a pretext for war because they represented a person's wealth.) Maeve was depicted as a vicious fighter and a voracious lover who once required seven men to sate her appetite. Her burial place is believed to be a huge mound of loose limestone, a Neolithic passage tomb that has never been thoroughly excavated, sitting on top of Mount Knocknarea. In height my Maeve is only ten feet taller than the typical twenty-five-foot *C. douglasii* that grows in the range of this species, but the circumference of her largest trunk is forty-one inches, more than three times the average, and the diameter of her crown is thirty-eight feet. (Maeve would be dwarfed by the tallest tree in the world, a coast redwood in California named Hyperion that soars to 379 feet and is 80 feet in circumference.) Maeve grows in the shadows of the big timber at Dark Acres, old 100-foot cottonwoods and 130-foot ponderosa pines. She's slightly shorter than the biggest recorded Douglas hawthorn in Idaho, and considerably smaller in circumference than what used to be the biggest Douglas hawthorn in the United States, which grew in Washington State's Beacon Rock State Park until 1993, when the Columbia River took out its roots and washed it into the Pacific.

I know Maeve is a record-setting giant because she was measured by an expert. One smoky August afternoon Helen Smith drove out to Dark Acres with an assistant, and we picked our way through a grove of hawthorns growing along the left bank of the Mabel, our widest slough, into a small clearing that serves as the courtyard of the Warrior Queen. Smith, a fire ecologist for the U.S. Forest Service who studies the positive and negative impact of wildfire and controlled burns on different kinds of forests, used a hypsometer to take Maeve's measure. This nifty instrument looks like a camcorder sheathed in a yellow case and costs around twenty-five hundred dollars. The recorder sights through an eyepiece at the object to be measured, then presses buttons that shoot lasers, which bounce off the target and feed the data they capture into an onboard computer. This crunches the numbers, after which they can be downloaded onto a laptop. Measuring Maeve would have been a simple matter if she had been standing alone in a pasture, like the first hawthorn I met on that memorable day. But because Maeve's crown and a couple of her trunks are tangled up with lesser hawthorns, it took a while to determine which was Maeve and which were the others. In fact, we were only guessing that Maeve was a *C. douglasii,* as opposed to one of the other species native to the region, such as *C. chrysocarpa* (fireberry) *C. succulenta* (fleshy hawthorn), *C. columbiana* (Columbia hawthorn), or a species that is relatively rare on the east side of the Continental Divide, *C. suksdorfii* (Suksdorf's hawthorn). She might even be a hybrid of these species. But one of Smith's protégés determined from the shape of the leaves and the ten pink stamens that she was indeed a *C. douglasii.*

Besides working as a researcher, Smith coordinates the Montana chapter of the National Champion Tree Program, sponsored by the oldest conservation organization in the country, American Forests. I found out about this project after I came across Maeve while foraging for morels. I could see how much bigger she was than her siblings and cousins, and decided to look for reports about other *C. douglasii* giants. That's when I came across the National Register of Big Trees, which sprang from a 1940 article in *American Forests* magazine, "Let's Find and Save the Biggest Trees." Private, corporate, and government landowners are encouraged to nominate big trees growing on their land as part of a public relations campaign to keep the reforestation of America on the economic and political agenda. Voters and consumers sometimes forget that forests have enormous economic value beyond the wood products they supply. Forests clean the air, harbor our watersheds, and provide the essentials that birds, insects, and wildlife need. Besides cheerleading for trees, American Forests has planted more than 46 million saplings since 1990 and intends to plant millions more in the coming years. These steps alone won't restore the vast canopies that existed in pre-Columbian America, but they will set an example and point the way.[1]

I was thrilled when American Forests sent me the news about Maeve. It turns out she's not just a big tree, she's the largest recorded Douglas hawthorn in the world. And she's in good company, as one of 751 grand champion trees, including 15 other species of hawthorn. She earned 86 points on the National Register's scale, compared to the score of 168 earned by a champion *C. monogyna* in Connecticut, and the 21 points awarded a reverchon hawthorn, *C. reverchonii,* in Dallas. I was intrigued to learn that there are forty-nine species of American hawthorn that currently have no national champion. Could I forge a new career finding and nominating the goliaths on this list? Maybe I could land a corporate sponsorship, say one of the big herbal supplement companies such as Wonder Labs or Puritan's Pride that market hawthorn products. Like a NASCAR driver I would show up at a landowner's door in a jumpsuit adorned with the company's logo calling myself the Nominator. In a sense I'd be like two fearless plant hunters, widely traveled botanists separated in time by 125 years, who scoured North America looking for new species to share with the world.

The first was David Douglas, the heroic and tragic Scottish botanist who found what would become known as the Douglas hawthorn in 1827 (of course, it had been known to indigenous peoples under different names for millennia).[2] Douglas was born near Perth in 1799, the son of a stonemason. He spent a few years in school, and at the age of eleven went to work for the head gardener at the estate of the earl of Mansfield at Scone Palace. After finishing his apprenticeship seven years later he went to work on the estate of Sir Robert Preston, tending to outdoor and greenhouse plants. An autodidact, he furthered his botanical and zoological education by reading every book on these subjects in Sir Robert's impressive library.

In 1820 Glasgow University hired Douglas to work in its botanical garden. It was here that he met a new professor of botany, William Hooker, who took Douglas into the field and taught him the art of pressing and drying plants, essential skills for the collector. Hooker recommended Douglas for a job with the Royal Horticultural Society, which was looking for a skilled botanist to send on an exploratory trip to America. Douglas would make three such trips. The first, in June 1823, took him from Liverpool to New York. From there he made excursions around the Northeast investigating North American trees, orchards, gardens, and birds, traveling as far west as the Detroit River. On an island in the river he collected seeds from the trees by blasting off their branches with his gun. These included at least two species of *Crataegus,* one of which he described as "of very large size" with huge fruit "almost like crab-apples." Pushing farther into Canada he found more hawthorns, which he believed were taller than those in the United States. In Albany he examined a farm whose owner, a former printer named Jesse Bull, had imported *C. oxycantha* (now *monogyna*) from Britain for the purpose of growing hedges to protect his gardens. Also during his explorations of the Northeast, Douglas examined plants from the Lewis and Clark Expedition that were flourishing in American gardens and collected seeds from rare species growing around Philadelphia, which he sent back to London. The quality of this material so impressed the Royal Horticultural Society that when the Hudson's Bay Company announced it would sponsor a collector to explore flora along the Columbia River in the Pacific Northwest, Douglas was the society's immediate choice. The work he would do for the society on this and other travels would earn him his reputation as the most courageous and accomplished botanical explorer in the New World. It eventually cost him his life.

Douglas boarded the *William and Ann* in July 1825. Nine months later, after sailing around Cape Horn at the southern tip of South America, he arrived at Fort Vancouver on the Columbia River in what is now Washington State. He went to work immediately.

As he began exploring along the big river Douglas found plants unknown in Europe and the settled parts of America almost every day he worked in the field. During the spring he collected flowers, in the summer leaves and branches, which he arranged to be sent back to London, and in the fall he returned to these same plants and collected seeds. The farther upstream he traveled the stranger the flora. One of the collections he sent home via ship in the autumn of 1826 included branches and needles of what he called "Oregon pine," which botanists later named the Douglas fir.

During the course of the year Douglas would range far and wide, exploring Oregon and Washington and becoming the first European to climb several peaks in the Cascade Range as part of his relentless quest to find new plants. Winter ended his journeys, but he used the time to copy his notes, press and dry plants, and ready them for shipment back to London. Although he was initially distrusted and feared by some of the Native peoples for his strange habits, such as wearing spectacles, drinking tea, and lighting his pipe by focusing the sun on it with a magnifying glass, when he became better known they nicknamed him "Grass Man."

On the first day of spring in 1826 he left Fort Vancouver accompanied by sixteen men in two boats and headed up the river. Their destination was Kettle Falls, 350 miles northeast, 30 miles south of what is now the Canadian border and 200 miles west of what is now Dark Acres. Douglas collected plants every step along the way. His first mention of what would be named *C. douglasii* in his honor by members of the Royal Horticulture Society was this entry in his journal, made on 13 May 1826, when the expedition halted at the junction of the Spokane River with the Columbia: "I met here Mr. John Work, with whom I was acquainted last year, and who sent me some valuable information about the plants and mountain sheep in this neighbourhood. I find that the package of seeds marked 'Wormwood of the Voyageurs' is *Tigarea tridentata;* that marked by myself as with a query is a very fine species of *Crataegus* found only in the interior."

On 24 May the expedition was encamped at Kettle Falls. Because its cascades cause spawning salmon to throw themselves from the water, this was an ancient and important fishing site for tribal people, who used nets on long poles to land their prey. Douglas wrote in his journal: "Showery, with a south-west wind. After turning some plants and taking others out of the presser, went out in quest of more." What he brought back that evening included a *Crataegus* that Douglas said was "the only one of the genus I have seen in the interior; on the edges of rivers and creeks; a low spreading shrub."

The *C. douglasii* seeds he sent back to London were planted. They sprouted and grew, revealing a previously unknown species. Most English people, accustomed to their large native hawthorns such as *C. monogyna,* would probably not be impressed by this mutt (none of the specimens was as tall as Maeve). But the scientists were thrilled.

The following spring the expedition made its way even farther up the Columbia into Canada. They crossed the Rocky Mountains and journeyed to Hudson Bay, where Douglas boarded a ship and returned to England. Two years later he was back in the Northwest collecting plants again. He headed south to explore the flora of California, finding five hundred new species. As a consequence of the fact that Oregon and Washington were owned by the British, and California was owned by Mexico, he was forced to spend two years there because it was politically more difficult to leave California than it was to enter it.

In 1833 Douglas sailed for Hawaii, then called the Sandwich Islands. He had wintered there several times before. The peculiar flora of the islands was a thrilling challenge and, of course, he climbed several mountains, in the company of his little terrier, Billy. In January 1834 he and a guide trekked up Mauna Kea, staying overnight in a lodge owned by two bullock hunters. About the bullocks, Douglas had written in his journal that "the grassy flanks of the mountain abound with wild cattle, the offspring of the stock left here by Captain Vancouver, and which now prove a very great benefit to this island." He had calculated that the volcano's summit was not eighteen thousand feet above the sea, as he had been led to believe, but only thirteen thousand (actually, it's 13,796).

Douglas returned to the volcano in July, intending to hike a trail skirting the slopes at the six-thousand-foot level. He was accompanied by Billy, a guide, and several porters. On 12 July, the morning of the hike, he stopped for breakfast at the home of Edward (Ned) Gurney, a bullock hunter, who agreed to accompany Douglas for a short distance down the road leading to the base of the volcano so that he could help the naturalist avoid the deep pits around certain waterholes where bullocks were known to drink. (The easiest way to kill a bullock, apparently, was to dig a big hole and hope the beast fell in, where it would be shot.) Gurney reported that before he parted with Douglas he warned him about three more pits a couple of miles ahead, two of them directly in the road, and the third to one side.

Later that morning two native people came across a gruesome sight. A bullock that had fallen into one of the pits was standing on the corpse of David Douglas. They ran to Gurney's house. The hunter grabbed his musket, and hurried to the pit, where he shot and killed the animal. They heard Billy barking, and found the dog a short distance away, along with Douglas' pack. It was speculated that Douglas had peered briefly into the pit, proceeded up the road, then, for some reason, left his dog and his pack, and returned to the pit for another look, when he somehow tripped and fell in. Although he had previously complained that his eyesight was failing and, in fact, was completely blind in one eye, almost no one believed that his death was an accident. For one thing, the ten gashes in Douglas' head were not consistent with the wounds produced by a goring bull. For another, he was reportedly carrying a purse full of gold coins, but it was never found. And Douglas was a skilled mountaineer who was no more likely to fall into a pit than eat belladonna seeds. It was also noted that Douglas' guide, Ned Gurney, was a convicted thief who had been deported to Botany Bay, Australia's notorious penal colony, where he spent seven years. Among the doubters of the accident theory it was speculated that Gurney had followed Douglas instead of returning home, killed the naturalist with an ax, stolen his purse, and dumped his body in a pit that had already trapped a bull.

Maybe. Probably.

As with so many investigations that fail to secure a confession or uncover a key piece of evidence, no one was prosecuted for the murder. Ned Gurney left Hawaii in 1839, and reportedly confessed to the crime on his deathbed. Billy was returned to England, where he spent the remainder of his life in the house of a parson. And for years after David Douglas died at the age of thirty-five, his collections of plants would continue to arrive by ship in London, and his technical descriptions of the 240 plants he introduced to Britain would continue to fill the pages of the *Transactions of the Royal Horticultural Society.*

The second botanist who has contributed mightily to our store of knowledge about the flora of North America is James Bird Phipps. Over the span of his long career he has published more than 60 professional papers about the communities, breeding systems, ecology, and identification of hawthorns. He has personally discovered and named twenty-seven new *Crataegus* species in North America. These publications are part of a body of work of more than 120 papers and 64 scientific lectures, a prodigious amount of research summarized in his beautifully illustrated *Hawthorns and Medlars.*[3] He is currently at work on the *Crataegus* section of the massive, 30-volume *Flora of North America,* which will include contributions from hundreds of specialists about every native plant on the continent—a vast improvement on the puny two-volume compendium on flora that was published in 1840.

Where did Phipps get his driving interest in this peculiar genus? He was born in Birmingham, England, in 1934. His father was a schoolteacher who came from a poor coal-mining family, but graduated with honors from Birmingham University with a degree in mathematical physics. Phipps' maternal grandfather was a commercial rose grower who built a grand house surrounded by gardens, orchards, and tennis courts. Phipps' parents were also talented gardeners, and they encouraged their son to develop his own small plot in one of the far corners of their lot in the city.[4]

The young Phipps was introduced to the natural world on walks around the countryside with his mother. His first contact with natural history came via the family's collection of cigarette cards, cardboard sheets used by tobacco companies such as John Player and Sons to reinforce the soft paper cigarette packages of the time, and as advertisements. A full-color illustration was printed on one side, and a brief caption on the other. The cards came in series of twenty-five or fifty and covered topics ranging from famous cricket players to trees and butterflies. From them Phipps learned that all plants and animals have two names, a common and a scientific name.

Phipps' serious passion for botany began when he was a teenager on family vacations and road trips to the hedged fields and tiny woodlands of the surrounding countryside. On one of these outings he came across a flower that was not in his guidebooks. He had the audacity to contact a famous orchid specialist at the Royal Botanical Gardens, Kew, who was so excited about the find he took the train up from London to see what turned out to be a southern marsh orchid, *Dactylorhiza praetermissa.* Phipps was accepted into the Botany School at Birmingham University. During college he learned to climb rocks and mountains, skills which served him well on what would become a lifelong quest for an understanding of the natural world. In 1954 he published his first scientific paper—regarding the discovery of the second British location of salsify, *Scorzonera humilis*—when he was still an undergraduate. This was the result of his work identifying and counting all the plant species in a "square" (square kilometer), as part of the Flora of Warwickshire project, a survey funded by an accomplished amateur botanist, Dorothy Cadbury, of the Cadbury chocolate family.

After graduating, in 1956 he boarded the *Pretoria Castle* for Cape Town, South Africa, where a train would take him fifteen hundred miles northeast to his first job, at the government herbarium in what was then Salisbury, Rhodesia (now Harare, Zimbabwe). The Rhodesian Ministry of Agriculture had hired him to explore, discover, identify, and write about plants that were new to Europeans. The train was so slow some passengers jumped off at one station to catch a movie, then hitched rides in automobiles to get to another station ahead of the train.

Phipps expected to arrive in a dusty colonial outpost surrounded by lion-infested savannahs. But 1956 Salisbury was a modern city of 250,000 laid out on a grid. The countryside was also much more varied than he had expected. Besides lions, it was home to hippos, Cape buffalos, wild dogs, hyenas, baboons, venomous snakes, and the number 1 carnivore, the crocodile. Malaria was common in some areas, and parts of the country were uninhabitable because of sleeping sickness borne by the tsetse fly. Other insect threats included African bees, driver ants, and termites, which ate the leather soles of Phipps' cricket boots. As he ventured into the field by Land Rover or on foot—safaris that lasted a week to three weeks—he learned the rules of the wild: never get lost, make sure you have a plan of retreat, look carefully before you move your hands or feet, and never drink water that hasn't been boiled.

After almost five years in Rhodesia, Phipps headed to the University of Western Ontario, where he wrote his doctoral thesis on African grasses, was awarded a Ph.D. in botany in 1969, and was hired to teach. He looked around for a new challenge, something that might yield interesting problems for a long time. "Lo and behold," he told me, "there was *Crataegus* with some twenty-five plus species in Ontario, many more in the rest of North America." Three aspects of this genus intrigued him: they were understudied, they were often difficult to identify, and it had been postulated that one of their methods of reproduction was apomixis: virgin births, a phenomenon that creates challenges for the taxonomist.

With the help of his graduate students, Phipps began collecting specimens throughout southern Ontario. Aided by a cytologist, or cell specialist, from India named Muniyamma, he conclusively demonstrated that hawthorns can reproduce by apomixis. To the taxonomist, slight genetic variations among these apomicts, repeated many times by cloning, can appear to be narrowly defined species and help make identifying apomictic genera a challenge. The other difficulty in the identification of hawthorns is their tendency to produce a prolific number of hybrids, more than twelve hundred in North America. This is called "the *Crataegus* problem."[5]

The naming of hawthorns, however, is the same as the naming of any group of plants—it is based on the possession of shared characteristics such as the number of stamens and the shapes of leaves. Each named species has a "type" specimen, which is the ultimate example of what constitutes the species. This is a dried and mounted flowering or fruiting twig (in the case of hawthorns) that is considered representative and is stored in a public herbarium, such as that of the Missouri Botanical Garden, the largest such collection in America. With apomictic groups of hawthorns botanists do not always agree, not so much on how many "species" there are—impossible to know because apomicts do not equate to normal sexual species—but rather on how many are realistically worth recognizing. Phipps' count of species runs to about 150 for North America.

Phipps has collected hawthorns in the forty-seven U.S. states that have significant populations, seven Canadian provinces and most of the high-altitude states of Mexico, publishing papers and giving lectures as he filled the herbarium at his university with some eleven thousand samples of *Crataegus,* some of which are duplicates that can be traded, like Phipps' boyhood cigarette cards, to other herbaria. "Phipps is the pre-eminent student of North American hawthorns and medlars," a fellow botanist wrote. "His energetic fieldwork and detailed revisionary studies have provided a wealth of new information about these plants."[6]

Students who are wondering whether they should become botanists might be intimidated by the notion that prolific collectors such as David Douglas and James Phipps have already discovered all the earth's plants. But the truth is, "new" plants are being named every year. In 2011, for example, collectors in Brazil discovered a little flower that stretches its branches to the ground to plant its own seeds, a trait known as geocarpy. (Because it is so dexterous they named it *Spigelia genuflexa.*)[7]

And the demand for new practical applications of the science is growing as a result of changes in the environment, facilitated by advances in technology. The countries of the world are slowly being forced to consider the implications of climate change on their agriculture and forestry. The increasing demand for natural, non-synthetic pharmaceuticals is driving a burgeoning industry for what are now called dietary supplements. And forensic botanists are even being called as expert witnesses in criminal trials. (Although at the time of this writing no *Crataegus* has figured in a courtroom drama, considering the strain of antic depravity that runs through human beings it will only be a matter of time.)

There is no better opportunity to explore the world than becoming a plant collector. "What can be more wonderful," Phipps told me, describing one of his field trips to Montana, "[than to stand] on a sunny day in mid-September, on the north-facing slopes of the isolated and remote Sweet Grass Hills, . . . gazing across the endless prairies of Alberta far below? The peace and calm of the scene is unforgettable and I have been blessed to see so much of nature in this way."

Now, back in the present, I was home for a couple of hours to feed the dogs and horses and pay some bills. I stared at my ringing cellphone. My hands began shaking. I had finally lowered the firewall that had been holding at bay the terrifying thought that I might have to face a life without Kitty. A week earlier I had taken her to the hospital after it was clear that the asthma she suffered from had turned into something much worse. Her doctor explained that the gray area on her X-rays was a serious infection in her lungs. The morning after she was admitted, both lungs collapsed. Her doctors induced a coma, intubated her lungs so they could be drained, and then re-inflated them. They took blood for the lab to analyze, hoping for a speedy diagnosis. Her family gathered with me around her bed, expecting the worst, and watching her vital signs on a monitor, especially the one that measured how much oxygen her lungs were taking in. The number slowly dropped. Meanwhile, I shifted my emotions into neutral, and filled my head with anything but the here and now.

Three days later the doctors finally had their diagnosis. It turned out to be acute eosinophilic pneumonia, a disease in which white blood cells called eosinophiles accumulate in the lungs, filling the air spaces where oxygen is extracted with their little dead bodies. In tropical countries this disease, which is relatively common, is caused by blood parasites that throw the body into a panic, producing armies of white cells to attack the invader. In the developed world it's relatively rare, and can be caused by cancer, drug abuse, medications, and exposure to certain irritants in the atmosphere—a firefighter developed the disease after inhalation of dust from the World Trade Center on 9/11, and it struck some U.S. soldiers stationed in the deserts of Iraq exposed for the first time to cigarettes and blowing sand. But in Kitty's case the doctors ruled out all these agents, including their prime suspect, the mold that grows on hay. It was "idiopathic," they said—cause unknown.

If the disease isn't too far advanced, high-dose injections of corticosteroids can cure it quickly. The nurses had begun dripping them into Kitty's blood this morning. But when I left the hospital eight hours later there had been no improvement in her oxygen intake.

I finally answered the phone. I listened for a moment, then hung up. Kitty's vital signs had improved so quickly in the past two hours her relieved doctors had decided that tomorrow they were going to bring her out of her coma. I leaned down and gave the dogs a kiss.

Then I put my arms around Maeve and held her like an old friend, because she was.

Epilogue

I can't look back any farther than the Middle Ages for illumination about my Irish ancestors because no evidence exists. But records show that the Morans were members of a *sept,* or a clan, belonging to a tribe in the western Ireland counties of Mayo and Sligo called *Uí Fiachrach*, whose central organization seems to have collapsed after the Norman invasion in 1066. A sept, from the Latin for "enclosure," might have its own chieftain, or more likely it would follow the lead of a more powerful family's strong man. The Morans claim a number of different crests, but they all bear three stars and the Latin words *Lucent in tenebris*—"They shine in darkness."

The nineteenth-century Morans of County Waterford didn't appear to shine at anything, however. Although many of the legal restraints holding down Catholics had been rescinded by the time my great-grandfather came of age, the Potato Famine and the enclosure of common land made it difficult for Thomas Moran to find work in Ireland. Compounding his lack of opportunity was the Encumbered Estates Act of 1849. This allowed land holdings whose owners were in heavy debt to be auctioned off. While 90 percent of some three thousand estates were bought by other Irish, this didn't help the Morans because the middlemen hired by native landlords were indifferent to peasants; they wanted them off the land so it could be used for cattle instead of crops. They jacked up the rents on tenant farmers and evicted nearly fifty thousand families between 1849 and 1854, filling the workhouses and flooding the island's already surplus labor force. In England, a similar land-use revolution forced peasants into the cities to work in industry. But in Ireland, with the exception of linen manufacture, there *was* no industry.[1]

So like a million of his countrymen, Thomas Moran fled to America. Following the exodus of his older brothers, he sailed to Boston in 1860 at the age of twenty-two with his mother, Honora Brigit, who had turned her back on Ireland, and apparently on her husband, Edmond, as well. In 1866 at the age of seventy, Edmond died of what was listed as chronic bronchitis but was more likely emphysema caused by a life of breathing the smoke of wood and peat. Thomas found work on a farm outside Chelsea, north of Boston, the only labor for which he was qualified.[2] The cascade of refugees overwhelmed Boston and prompted a strong anti-Catholic and anti-Irish sentiment among old Yankee families, who could recite their lineage to the *Mayflower* as fluidly as they could the pedigrees of their horses. For Thomas, conditions were no better than they had been in Waterford. It was just as well he could not read, because the infamous sign "No Irish Need Apply" hung everywhere from shops and factories.

In 1861 he went downtown to sign up for the Civil War but the recruiters told him that the North had enough soldiers and would win the war in a few weeks. In a huff, he sailed off for San Francisco by way of Panama, walking across the Isthmus to the Pacific. By now the gold rush was long over, and the gold was gone. He milked cows in Santa Clara, California, sleeping with his hands in pails of cold water because they were so swollen from the work.

Then, with a friend and a horse, he left California and headed north for the gold fields of Montana, where new strikes at Bannack and Last Chance Gulch were drawing prospectors from all over the world. Employing the "ride-and-tie" method, in which one man rode ahead, and tied the animal to a tree for the next man to mount when he caught up on foot, they reached Umatilla, Oregon, where they parted company.

Along the way Moran passed through great swaths of hawthorn thickets stretching from northern California to Portland and along the Columbia River winding through Washington. These were the *C. douglasii* named after David Douglas, and the Columbia hawthorn, *C. columbiana,* both mentioned in the journals of Lewis and Clark. Pushing on alone with a pack mule, Moran traveled west along the Columbia. A detachment of soldiers from Fort Walla Walla gave him a horse and persuaded him to hole up for the winter inside the fort because of marauding Nez Perce and Shoshone. On the way to the fort the column was attacked by Indians; when Moran removed his bedroll from his saddle he found a bullet lodged inside.

While wandering around in the Highwood Mountains and along Last Chance Gulch in the present-day town of Helena, Moran carried a Sharps .50 caliber needle gun issued to him by the army for protection against the Indians. As the Territory's tribes were rounded up and herded onto the reservations to make way for the miners and the cattlemen who came to feed the miners, I wonder whether he grasped the irony—the landless and oppressed Irishman was now aiding forces that aimed to confine Montana's natives, suppress their language, deprive them of their religion, and take their land.

In January 1866, at the age of twenty-eight, Moran finally found his place in the world. It was an accident, really, and one that almost killed him. When a bogus rumor began circulating about a big gold strike in the Sun River Country, he joined a mob of ill-clad miners and rushed a hundred miles north from what is now Helena on foot and on horseback. No one reported finding any gold. The chinook that bore this rumor was followed by an arctic storm, which killed some and cost scores of others various body parts. Moran headed back to Helena in a blizzard. Many years later he told a frontier chronicler that he had camped with seven hundred men at St. Peter's Mission in temperatures that reached −40 degrees.[3] The Jesuit fathers at the mission did their best to keep the ragged crew alive. Grateful for his life and impressed by their generosity, Moran never left the mission.

When the Blackfeet went on the warpath that spring, killing settlers, burning buildings, and slaughtering cattle, the Jesuits abandoned the mission, leaving Moran as caretaker, along with Edward Lewis, who would become his next-door neighbor and life-long friend. The two kept their scalps, primarily because Edward had wisely married the daughter of Heavy Shield, a Blackfeet chief. When the hostilities were over, Moran set about realizing his quest for the two things that had been denied him in Ireland: education and land. As late as 1870, when one of the Jesuits wrote a letter for him, he was still illiterate. But fifteen years later he must have learned how to read because he was appointed the postmaster for St. Peter's. He acquired more than seventeen hundred acres in the foothills and valleys around the mission—land once swarming with bison; land that had belonged to the Blackfeet—and became one of the young state's most successful farmers. Although the Jesuits had visited the mission after they abandoned it, St. Peter's wasn't reopened until 1874.[4]

By today's standards the value of the cash Moran paid for his first quarter section in 1879 would be around twenty thousand dollars. Even in the inflated economy of the Montana Territory a decade before statehood, where it cost a dollar to send a letter, this was a sizable sum of disposable wealth: an Irish farm laborer was likely to possess such an amount only if he'd robbed a bank or struck gold. By all accounts Moran was a thoroughly law-abiding citizen. He served on posses chasing murderers and horse thieves. It is not known how Moran acquired this sum, but a favorite family speculation holds that he discovered a cache of gold stolen from miners and stagecoaches by the notorious sheriff and outlaw Henry Plummer, who was hanged in 1864 by vigilantes. This cache was reputedly buried near St. Peter's, near a geological formation known as Bird Tail Rock.[5]

It was at St. Peter's in 1866 that the paths of Thomas Moran and Thomas Francis Meagher crossed once more. For his part in the Rebellion of 1848, Meagher had been exiled to Van Diemen's Land, now Tasmania, after his death sentence for sedition was commuted. In 1852 he escaped and ended up in New York City. Unlike Moran, he was not only accepted into the U.S. Army, he was commissioned brigadier general and given command of the Irish Brigade. In 1865 he was appointed military secretary of the Montana Territory. When he arrived in Helena the governor promptly resigned, making Meagher acting governor, as well. That autumn, on a trip across the Territory to negotiate with the Blackfeet, he spent several nights at St. Peter's with his entourage, waiting out a storm. After the Jesuits abandoned the mission the next year, Meagher arranged to have it rented to the U.S. Army. On 1 July 1867 he fell or was pushed from a steamboat into the Missouri River. The circumstances were mysterious, and his body was never found. An enormous statue of him was erected in front of the Montana Capitol in Helena. As in the smaller statue in Waterford City, Meagher is on a horse, sword raised, ready to charge.

I was middle-aged before I first laid eyes on St. Peter's, even though my grandfather had been born there and my mother, as well. The valley where its ruins lie are in the foothills of the Rockies less than two hundred miles from Dark Acres. My father and my mother's family never spoke of the place in my presence, and I never gave it a thought until I began allowing myself to think dispassionately about her death. The day of my autumn visit was sunny, windless and warm. I poked around the crumbling, scorched walls of the two enormous stone dormitories that had been built by the Jesuits and Ursuline nuns to house Blackfeet children while the paganism was bleached from their souls. The Blackrobes must have thought these buildings would last forever. But the boys' dorm burned to the ground one winter and, a few years later the girls' as well. I wandered across a cattle pasture to the little whitewashed church Moran helped the Jesuits build from rough, squared cottonwood timbers chinked with clay. Until recently, my aunts and uncles had celebrated Christmas mass there every year. I peered through the windows at the statues of Saint Peter and the Madonna and Child. It was so quiet I thought something was wrong with my hearing.

I walked from the church to the fenced graveyard at the top of a rise and found Thomas Moran's headstone. The elements had almost erased his name. A badger had bored its den straight down into his tomb. I thought of the photograph of him, published in a history of St. Peter's, showing a tall, lanky, man with thinning hair, sporting a grizzled white beard, wearing bib overalls and a work shirt, propping himself up with a cane. Like everyone in our family he had big ears and a prominent, Roman nose.

Back in my truck I drove a couple of miles back down Mission Road, parked, and began walking on land he once owned toward the barn he had built. I came to a finger creek hedged by twenty-foot trees crowned with a riot of yellow fall leaves, swarming with birds feasting on bright red fruit, which had grown so fat the branches were fairly groaning under its weight. The irony was seamless. Armed with vicious two-inch spikes, these were hawthorns.

Notes

ONE

The World's Busiest Tree

1. Kate Waters, *On the "Mayflower": Voyage of the Ship's Apprentice and a Passenger Girl (*New York: Scholastic, 1999), 36.

2. Sylvia Plath, "The Bee Meeting," in Plath, *Ariel* (New York: Faber and Faber, 1965), 81.

3. J. K. Rowling, *Harry Potter and the Sorcerer's Stone* (Pottermore Limited, Kindle edition, Chapter 5, 2012); "Wand Woods," www.pottermore.wikia.com/wiki/Wand Woods.

4. Marcel Proust, *Swann's Way, In Search of Lost Time,* vol. 1 (New York: Modern Library, Kindle edition, first published in 1913).

TWO

Under the Hawthorn Tree

Epigraph: John Barrow, *A Tour Round Ireland, Through the Sea-coast Counties, in the Autumn of 1835* (London: J. Murray, 1836), 246.

1. See Christine Zucchelli, *Trees of Inspiration: Sacred Trees and Bushes of Ireland* (Cork, Ireland: Collins, 2009); Herbert Thurston, "Devotion to the Blessed Virgin Mary," in *The Catholic Encyclopedia,* vol. 15 (New York: Appleton, 1912).

2. Sue Moran, "Thomas Moran" (2004), unpublished manuscript; copy in the author's possession. Investigations of a family's Irish roots often rely on educated guesswork. In her excellent biographical sketch of our great-grandfather, my cousin Sue Moran unearthed numerous documents and accounts of his adventures in America. But as she says, "Little is known of Thomas' early life in Ireland." The Morans are not listed in any census because most of those records were lost in a fire in 1922 that destroyed the Public Records Office in Dublin during the Irish civil war. But many peasants were included in Griffith's Valuation, a survey of every plot of land in Ireland that also lists the names of the landlord and the tenants, although only the household heads were counted. Carried out between 1847 and 1864—the

valuation for County Waterford was released in 1853—Griffith's was an assessment of land values that was used to determine the amount of various taxes levied on landowners, one of which was used to fund the notorious workhouses that were built as a result of the 1838 Poor Relief Act. Tenant farmers, large and small, were listed as the "occupiers" of plots of land they rented. As head of the Moran family, Edmond would have been listed in Griffith's if he had owned or rented land, no matter how small the plot. But his name does not appear. Perhaps the Morans are listed on the rolls of the paupers admitted to the workhouse in Carrick-on-Suir between 1838 and 1860. But no one I talked with in Ireland knew what happened to these records. People avoided the workhouse, entering only when they had run out of options; it is described as "the most feared and hated institution ever established in Ireland" by John O'Connor in *The Workhouses of Ireland* (Dublin: Anvil, 1995).

However, there are two surviving documents that frame what is probably the life of my great-grandfather in Ireland. The first is his birth certificate. Because it was filed in the parish of Mothel and Rathgormack his family probably lived in this rural area. But where? Irish peasants rarely moved around the country and tended to die close to where they were born, and Edmond's death certificate, filed in 1866, suggests that the family lived in the township of Ballynacurra a mile or so from the village of Rathgormack and about the same distance from what is left of the village of Mothel. His death certificate also explains why he had no entry in Griffith's—his occupation is listed as "laborer." Griffith's didn't include farm laborers because they usually lived on land owned and rented by someone else, trading their sweat for a residence. There is a remote possibility that the Morans were squatting on wasteland—a swamp or a rocky slope, for example—but by the time of the famine there was little of even this land that had not been enclosed.

I have therefore painted a broad account of the Morans in Ireland based on the lives of most small-tenant farmers and farm workers in that part of County Waterford in the middle of the nineteenth century.

3. See William Williams, *Creating Irish Tourism: The First Century, 1750–1850* (London: Anthem, 2011), 178–79.

4. See Patrick C. Power, *Carrick-on-Suir: Town and District, 1800–2000* (Carrick-on-Suir: Carrick Books, 2003). Much of my description of life in north-central County Waterford in the first half of the nineteenth century comes from Power, who used many primary sources, family stories, and newspapers.

5. Henry D. Inglis, *A Journey Throughout Ireland During the Spring, Summer and Autumn of 1834*, vol. 1 (London: Whittaker, 1835), 72–73.

6. On the potato, see Thomas Keneally, *The Great Shame* (New York: Doubleday, 1998), 8.

7. Keneally, *Great Shame*, 9; see also Cormac O Gráda, *Ireland Before and After the Famine: Explorations in Economic History, 1800–1925* (Manchester: Manchester University Press, 1993).

8. See Bob Curran, *The Truth About the Leprechaun* (Dublin: Wolfhound, 2000), 41.

9. Katharine Briggs, "Changelings," *An Encyclopedia of Fairies, Hobgoblins,*

Brownies, Boogies, and Other Supernatural Creatures (New York: Pantheon, 1976), 71.

10. John Bellamy Foster, "Malthus' Essay on Population at Age 200," *Monthly Review* 50, no. 7 (December 1998); Power, *Carrick-on-Suir,* 85.

11. Power, *Carrick-on-Suir,* 82; Susan Allport, *The Primal Feast: Food, Sex, Foraging, and Love* (Bloomington: iUniverse, 2003), 17.

12. Thefts are reported in *Tipperary Free Press,* 17 July 1847.

13. See Power, *Carrick-on-Suir.*

14. Marita Conlon-McKenna, *Under the Hawthorn Tree* (Dublin: O'Brien, 1990).

15. See Michael J. Winstanley, *Ireland and the Land Question, 1800–1922* (London: Routledge, Kindle edition, first published in 1994); D. J. Hickey and J. E. Doherty, *A Dictionary of Irish History* (Dublin: Gill and Macmillan, 1980), 86; Virginia Yans-McLaughlin, *Immigration Reconsidered: History, Sociology, and Politics* (Oxford: Oxford University Press, 1990), 100.

16. See Julian Hoppitt, *Parliaments, Nations and Identities in Britain and Ireland, 1660–1850* (Manchester, U.K.: Manchester University Press, 2003), 85–102; Oliver Rackham, *The History of the Countryside* (London: Weidenfeld and Nicolson, 1987), 190; Grenville Astill and Annie Grant, eds., *The Countryside of Medieval England* (Oxford: Blackwell, 1988), 23 and 64.

17. See Allan Kulikoff, *From British Peasants to Colonial American Farmers* (Chapel Hill: University of North Carolina Press, 2000), 11; Enclosure Act of 1845, at the National Archives Web site, Legislation.gov.uk; Hoppitt, *Parliaments, Nations and Identities,* 94.

18. James S. Donnelly, *Irish Agrarian Rebellion: The Whiteboys of 1769–76* (Dublin: Royal Irish Academy, 1983).

19. Richard P. Davis, *The Young Ireland Movement* (Dublin: Gill and Macmillan, 1988).

20. See Power, *Carrick-on-Suir.*

21. Keneally, *Great Shame,* 179.

22. Explanation of the statue is from the sculptor, Catherine Greene, via e-mail to the author, 21 February 2014.

THREE
The Celtic Forge

1. *The Catholic Encyclopedia,* vol. 13 (New York: Encyclopedia Press, 1913), 126; for an extensive discussion of the Penal Laws see Patrick Francis Moran, *The Catholics of Ireland Under the Penal Laws in the Eighteenth Century* (London: Catholic Truth Society, 1900).

2. See Padraig Lenihan, *1690: Battle of the Boyne* (Gloucestershire, U.K.: Tempus, 2003); Thomas Keneally, *The Great Shame* (New York: Doubleday, 1998), 8. The tale of Burke's abrupt departure is a favorite family story.

3. Antonia McManus, *The Irish Hedge School and Its Books: 1695–1831* (Dublin: Four Courts, 2004); Michael O'Laughlin, "I Never Carried the Sod," an Irish

hedge school history, *Irish Central,* 10 October 2009, available at http://www.irishcentral.com.

4. Joel Mokyr, *Why Ireland Starved: A Quantitative and Analytical History of the Irish Economy, 1800–1850* (Oxford: Routledge, 2013), 184.

5. See Barry Raftery, *Pagan Celtic Ireland: The Enigma of the Irish Iron Age* (London: Thames and Hudson, 1994); Simon James, *The Atlantic Celts: Ancient People or Modern Invention?* (Madison: University of Wisconsin Press, 1999), 136.

6. Tim Murray, *Milestones in Archaeology* (Santa Barbara, Calif.: ABC-CLIO, 2007), 221; Geoffrey of Monmouth, *History of the Kings of Britain,* 1138; Paul Jacobsthal, *Early Celtic Art: Plates* (Oxford: Clarendon, 1969).

7. For more about Celts and their cult of the head see Barry Cunliffe, *The Ancient Celts* (Oxford: Oxford University Press, 1997); Anne Ross, *Pagan Celtic Britain: Studies in Iconography and Tradition* (London: Sphere, 1974), 161–62; Miranda Green, *The Gods of the Celts* (Gloucestershire, U.K.: Sutton, 2004); Ronald Hutton, *The Pagan Religions of the Ancient British Isles: Their Nature and Legacy* (Oxford: Blackwell, 1991), 195.

8. Anthony M. Snodgrass, *The Dark Age of Greece* (Edinburgh: Edinburgh University Press, 2000), 286–87; F. B. Vagn, "Iron—a Very Special Metal," in *Iron and Steel in Ancient Times* (Copenhagen: KDV Selskab, 2005), 63–85.

9. Ernestina Badal Garcia, Yolanda Carrión Marco, Miguel Macías Enguídanos, and María Ntinou, *Wood and Charcoal: Evidence for Human and Natural History* (Valencia, Spain: Departament de Prehistòria i Arqueologia, Universitat de València, 2012), 233; Della Hooke, *Trees in Anglo-Saxon England: Literature, Lore and Landscape* (Woodbridge, U.K.: Boydell and Brewer, 2010), 143; "Archaeological Excavation Report, E3826, Catherweelder 7, County Galway, Iron Working Site," *Eachtra Journal* 8 (October 2010): 8.

10. Stephen Allen, *Celtic Warrior: 300 BC–AD 100* (Oxford: Osprey, 2001), 46.

11. Peter Berresford Ellis, *The Celts: A History* (New York: Carroll and Graf, 2004).

12. See Titus Livy, *The Early History of Rome,* trans. Aubrey de Sélincourt (New York: Penguin, 2002) book 5, chap. 37.

13. Ibid.

14. See Vagn, *Iron and Steel,* 17.

15. D. Killick and R. B. Gordon, "The Mechanism of Iron Production in the Bloomery Furnace," in *Proceedings of the 26th International Archaeometry Symposium, Held at University of Toronto, Toronto, Canada, May 16th to May 20th 1988,* ed. R. M. Farquhar, R. G. V. Hancock, and L. A. Pavlish (Toronto: Archaeometry Laboratory, Department of Physics, University of Toronto, 1989), 120–23.

FOUR
The Hedge Layers

1. William J. Curtis, "Events in the Life of Curtis G. Culin 3rd" (2001), posted at https://groups.yahoo.com/neo/groups/G104/conversations/topics/99.

2. The film can be seen at www.youtube.com/watch?v=RsRmEcKrYUE;

Frank Carlone, "History of the Essex Troop," posted at http://newarkmilitary.com/essextroop/historycarlone.php.

3. The photograph is reproduced in Robert Fridlington and Lawrence Fuhro, *Cranford* (Mount Pleasant, S.C.: Arcadia, 1995), 40.

4. Leo Dougherty, *The Battle of the Hedgerows: Bradley's First Army in Normandy, June–July 1944* (St. Paul: MBI, 2001), 202.

5. Harold J. Samsel, *Operational History of the 102nd Cavalry Regiment* (self-published, date unknown).

6. Robert Wolton, Nigel Adams, Emily Ledder, et al., "Report on the Hedgelink Visit to the Hedges and Orchards of Normandy, France" (May 2010), http://www.hedgelink.org.uk/european-hedges.htm.

7. Martin Blumenson and the editors of Time-Life Books, *Liberation* (New York: Time-Life, 1978), 20; P. Brunet, *Les Bocages, histoire, écologie, économie* (Rennes, France: Université de Rennes, 1977), 37–41; Martin Blumenson, *Breakout and Pursuit* (Washington D.C.: Center of Military History, 1993), 11.

8. Julius Caesar, *Caesar's Commentaries: On the Gallic War and on the Civil War* (Project Gutenberg, online version).

9. Bradley, quoted in Russell F. Weigley, *Eisenhower's Lieutenants* (Bloomington: Indiana University Press, 1981), 98.

10. "Ryan's Slaughter," *Independent.ie* (online edition, 10 May 2007).

11. Hal Boyle, "Battlefront 'Invention'—Tale of Sergeant Who Whipped Hedgerows," *Sandusky Register,* 16 December 1948, 6.

12. See Michel Rouche, "Private Life Conquers State and Society," in *A History of Private Life,* ed. Paul Veyne (Cambridge: Harvard University Press, 1987), 1:428; John C. McManus, *The Americans at Normandy: The Summer of 1944—The American War from the Normandy Beaches to Falaise* (New York: Macmillan, 2005), 97.

13. Dwight D. Eisenhower: "Remarks upon Receiving the Hoover Medal Award," 10 January 1961. Available at the American Presidency Project, http://www.presidency.ucsb.edu/ws/?pid=12068.

14. See Steven J. Zaloga, "Normandy Legends: The Culin Hedgerow Cutter," 1 July 2001), at the Osprey Publishing Web site: www.ospreypublishing.com/articles/world_war_2/normandy_legends.

15. See Hugh Barker, *Hedge Britannia: A Curious History of a British Obsession* (London: Bloomsbury, Kindle edition, 2012).

16. Christopher Long, "Hedge-Coppicing and Hedge-Laying in the Bocage Virois" (February 2011), www.christopherlong.co.uk/oth/hedges.html.

17. Max D. Hooper, Ernest Pollard, and Norman Winfrid Moore, *Hedges* (Glasgow: Collins, 1974).

18. A few good sources to help the hedge layer create a stock-proof hedge are Murray MacLean, *Hedges and Hedgelaying* (Marlborough, U.K.: Crowood, 2006); Marius de Geus, *Fences and Freedom: The Philosophy of Hedgelaying* (Dublin: International Books, 2003); Valeria Greaves, *Hedgelaying Explained* (Devon, U.K.: National Hedgelaying Society, 2002).

19. James D. Mauseth, *Botany: An Introduction to Plant Biology* (Sudbury, Mass.: Jones and Bartlett, 2012), 343–51.

20. For more on hedge-laying practices in England and France, see the National Hedgelaying Society Web site, www.hedgelaying.org.uk/.

21. Neil Foulkes, e-mails to the author (29 August–4 September 2013).

22. "Barley Farming in the UK," UK Agriculture, www.ukagriculture.com/crops/barley_uk.cfm; Bill Bryson, *Notes from a Small Island* (London: Transworld, 2010, Kindle edition); Alison Healy, "Survey Finds Most Hedgerows Will Die If They Are Not Actively Managed," *Irish Times,* 15 January 2014, online edition.

23. See Barker, *Hedge Britannia.*

24. For figures, see Robert Wolton, Nigel Adams, Emily Ledder, et al., "Hedgelink Visit to the Hedges and Orchards of Normandy, France," 8–10 May 2010, available at http://www.hedgelink.org.uk/files/Hedgelink%20-%20Report%20Visit%20to%20Normandy%20May%202010.pdf.

25. Thomas Karl et al., "Efficient Atmospheric Cleansing of Oxidized Organic Trace Gases by Vegetation," *Science,* 5 November 2010, 816–19.

26. Karl's comments about isoprene and monoterpenes in an e-mail to the author, 17 October 2014. For Karl's comments about the cleansing ability of trees, see "Plants Play Larger Role Than Thought in Cleaning up Air Pollution," NCAR UCAR AtmosNews, 21 October, 2010, available at www2.ucar.edu/atmosnews/news/2937/plants-play-larger-role-thought-cleaning-air-pollution. For Karl's comments about the "right" and "wrong" trees see Bruce Finley, "Deciduous Trees Have Decidedly Beneficial Impact on Air Pollution," *Denver Post,* online edition, posted 22 October 2010. To compare rates of VOC emissions see the United States Forest Service report "Estimated Biogenic VOC Emission Rates for Common U.S. Trees and Shrubs," http://www.nrs.fs.fed.us/units/urban/local-resources/downloads/vocrates.pdf.

27. Emine Sinmaz, "Charles and His Coat of Many Patches: Prince Shows off His Trusted Old Gardening Jacket, Complete with Holes and Tears, During Special Edition of Countryfile," *Mail Online,* 10 March 2013, available at http://www.dailymail.co.uk/news/article-2291333/Countryfile-Prince-Charles-shows-coat-patches-complete-holes-tears.html

28. See W. Gordon Bonn and T. van der Zwet, "Distribution and Economic Importance of Fire Blight," in *Fire Blight: The Disease and Its Causative Agent, Erwinia amylovora,* ed. J. L. Vanneste (Wallingford, U.K.: CABI, 2000).

29. Louisa Anne Meredith, *My Home in Tasmania: During a Residence of Nine Years,* vol. 1 (Cambridge: Cambridge University Press, 2010).

30. Fiona Blackwood, "Hedge Man," transcript of Stateline broadcast (12 October 2004), available at http://www.abc.net.au/stateline/tas/content/2003/s1262651.htm; Darrel Odgers, *Tasmania: A Guide* (Cameray, Australia: Kangaroo, 1989), 92.

31. James Boxhall, e-mail to the author regarding his work and that of Ellis, 17 September 2013.

32. Mark Lewis, *Cane Toads: An Unnatural History,* documentary (Film Australia, 1988).

33. For more about hedges and hedge plants in Australia and New Zealand see Larry Price, "Hedges and Shelterbelts on the Canterbury Plains, New Zealand: Transformation of the Antipodean Landscape," *Annals of the Association of American Geographers* 83, no. 1 (March 1993): 119–40; F. N. Howes, "Fence and Barrier Plants in Warm Climates," *Kew Bulletin* 1, no. 2 (1946): 51–87.

34. Roy Moxham, *The Great Hedge of India* (New York: Carroll and Graf, 2001).

35. Ibid.

FIVE
The American Thorn

1. For more about Washington the farmer see Alan Fusonie, *George Washington: Pioneer Farmer* (Charlottesville: University of Virginia Press, 1998); Robert F. Dalzell and Lee Baldwin Dalzell, *George Washington's Mount Vernon: At Home in Revolutionary America* (Oxford: Oxford University Press, 2000).

2. See Allan Greenberg, *George Washington, Architect* (Winterbourne, U.K.: Papadakis, 1999).

3. John C. Fitzpatrick, ed., *The Writings of George Washington from the Original Manuscript Sources, 1745–1799,* vol. 35: *March 30, 1796–July 31, 1797* (Washington. D.C.: United States Government Printing Office, 1939), 66.

4. For a description of invasive plants in the Pacific Northwest see Washington State Nursery and Landscape Association et al., *Garden Wise* (Federal Way, Wash., 2013), available at http://www.nwcb.wa.gov/publications/western_garden_wise.pdf. For an account of Washington's order of whitethorns see Paul Leland Haworth, *George Washington: Farmer* (Brooklyn, N.Y.: Bobbs-Merrill, 1915), 163.

5. Fitzpatrick, *Writings of George Washington,* 181–82.

6. See Charles Sprague Sargent, "The Trees at Mount Vernon," *Annual Report of the Mount Vernon Ladies' Association of the Union* (1917), 30.

7. See Avery Craven, *Soil Exhaustion as a Factor in the Agricultural History of Virginia and Maryland, 1606–1860* (Columbia: University of South Carolina Press, 2006), 110.

8. See Edwin Morris Betts, *Thomas Jefferson's Garden Book, 1766–1824* (Whitefish, Mont.: Kessinger, 2010).

9. Frederick Doveton Nichols, *Thomas Jefferson: Landscape Architect* (Charlottesville: University of Virginia Press), 143.

10. Betts, *Jefferson's Garden Book.*

11. Edmund Quincy, *Life of Josiah Quincy of Massachusetts* (Boston: Ticknor and Fields, 1867).

12. "Hawthorn Hedges in New England," *New England Farmer* 1, no. 1 (August 1822): 2–3; Quincy, *Life of Josiah Quincy,* 367.

13. Spencer C. Tucker, *The Encyclopedia of the War of 1812* (Santa Barbara, Calif.: ABC-CLIO, 2012), 849.

14. Quincy, *Life of Josiah Quincy,* 366.

15. William Cobbett, *The American Gardener; or, A Treatise on the Situation,*

Soil, Fencing, and Laying Out of Gardens; on the Making and Managing of Hot-beds and Green-houses; and on the Propagation and Cultivation of the Several Sorts of Vegetables, Herbs, Fruits, and Flowers (New York: Turner and Hayden, 1846), 25.

16. Ibid., 31.

17. Ibid., 28.

18. Michael Kammen, *Digging Up the Dead: A History of Notable American Reburials* (Chicago: University of Chicago Press, 2010), 74–77.

19. "Hedging," *American Agriculturist,* vol. 9 (1851): 298.

20. "A Chapter on Hedges," *Horticulturist and Journal of Rural Art and Rural Taste* 1, no. 8 (February 1847): 345–55.

21. See Donald Culrose Peattie, *A Natural History of Western Trees* (Boston: Houghton Mifflin, 1950), 477–82.

22. Carriel Mary Turner, *The Life of Jonathan Baldwin Turner* (Stockbridge, Mass.: Hard Press, 1911).

23. See James Edward David, *Frontier Illinois* (Bloomington: Indiana University Press, 2000).

24. *The Western Agriculturist, and Practical Farmer's Guide* (Robinson and Fairbanks, 1830, Google ebook), 53.

25. See Turner, *Life of Jonathan Baldwin Turner,* 65.

26. Alan W. Corson, "Planting Hedges," *Pennsylvania Farm Journal* 1 (1852): 207.

27. "Winterthur Bloom Report No. 20" (Winterthur, Del.: Winterthur, 29 May 2013), 2.

28. Laws of the State of Delaware, 19, Part 1 (1891); Records from the United State Patent and Trademark Office.

SIX
The Return of the Native

1. Susan Olp, "Tribal Members Work to Preserve Their Language," *News from Indian Country* (July 2012), http://www.indiancountrynews.com/index.php/news/279-culture/12573-tribal-members-work-to-preserve-their-language; Alma Hogan Snell, *A Taste of Heritage: Crow Indian Recipes and Herbal Medicines* (Lincoln: University of Nebraska Press, 2006).

2. Tim McCleary e-mail to the author, 26 November 2011; Frank B. Linderman, *Pretty-Shield: Medicine Woman of the Crows* (New York: Harper Collins, 1932).

3. Snell, *Taste of Heritage,* 135.

4. Ibid., 135.

5. Richard Mabey, *Food for Free* (Dublin: Collins, 1972, Kindle edition); "Springtime's Foraging Treats," *The Guardian,* 5 January 2007.

6. Snell, *Taste of Heritage,* xxi.

7. Michael Gard, *The Obesity Epidemic: Science, Morality and Ideology* (London: Routledge, 2004).

8. John R. Speakman, "Thrifty Genes for Obesity and the Metabolic Syn-

drome—Time to Call off the Search?" *Diabetes and Vascular Disease Research* 3, no. 1 (May 2006): 7–11.

9. John R. Speakman, "Thrifty Genes for Obesity and Diabetes, an Attractive but Flawed Idea and an Alternative Scenario: The 'Drifty Gene' Hypothesis," *International Journal of Obesity* 32 (2008): 1611–17.

10. For diabetes rates among American Indians, Montanans generally, and the Crow in particular see http://healthinfo.montana.edu/County%20Profiles/Big%20Horn%20County.pdf and http://www.cdc.gov/diabetes/pubs/pdf/diabetesreportcard.pdf. Snell, *Taste of Heritage*, 4.

11. James Mooney, *The Ghost Dance Religion and the Sioux Outbreak of 1890* (1896; Google eBook), 996 and 1014.

12. John C. Hellson, *Ethnobotany of the Blackfoot Indians* (Ottawa: National Museums of Canada, 1974).

13. Rosalyn LaPier, "Blackfeet Botanist: Annie Mad Plume Wall," *Montana Naturalist* (Fall 2005), 4–5; Blackfeet Community College and the University of Arizona, "Native Plants and Nutrition," http://www.nptao.arizona.edu/pdf/BlackfeetFinalReport.pdf; telephone conversation with Rosalyn LaPier, 10 October 2005.

14. For a comprehensive study of the plants used by North American Indians see Daniel E. Moerman, *Native American Ethnobotany* (Portland, Ore.: Timber, 1998).

15. See Mary Beth Trubitt, "The Production and Exchange of Marine Shell Prestige Goods," *Journal of Archaeological Research* 11, no. 3 (September 2003): 243–77.

16. Charles A. Geyer, "Notes on the Vegetation and General Character of the Missouri and Oregon Territories During the Years 1843 and 1844," *London Journal of Botany* 5 (1846): 300.

17. All the information about how Bittl makes a bow is from Michael Bittl, "Bows Made of Roses Vol. 1, Beauty All Around" (22 June 2011), posted at *Bow Explosion*, redhawk55.wordpress.com/tag/holzbogen/.

18. See James W. Herrick, *Iroquois Medical Botany* (Syracuse: Syracuse University Press, 1997), 161–62.

19. Robert M. Utley, *Cavalier in Buckskin* (Norman: University of Oklahoma Press, 1988), 192–93.

SEVEN
The Tree of Heroes

1. J. A. G. Roberts, *A Concise History of China* (Cambridge: Harvard University Press, 1999).

2. See Paul Clark, *The Chinese Cultural Revolution: A History* (Cambridge: Cambridge University Press, 2008).

3. Ai Mi, *Under the Hawthorn Tree* (London: Virago Press, 2012).

4. Patrick Frater, "Zhang Yimou to Be Fined $1.3 million for Having Too Many Kids," *Variety*, 29 December 2013.

5. The information about Jiahu is taken from Barbara Li Smith and Yun Kuen Lee, "Mortuary Treatment, Pathology and Social Relations of the Jiahu Community," *Asian Perspectives* 47 (2008): 242–98, and Juzhong Zhang, Garman Harbottle, Changsui Wang, and Zhaochen Kong, "Oldest Playable Musical Instruments Found at Jiahu Early Neolithic Site in China," *Nature,* September 1999, 366–68.

6. Patrick E. McGovern, *Uncorking the Past: The Quest for Wine, Beer and Other Alcoholic Beverages* (Berkeley: University of California Press, Kindle edition, 2010).

7. See James Bird Phipps, *Hawthorns and Medlars* (Portland, Ore.: Timber Press, 2003), 90.

8. See McGovern, *Uncorking the Past,* also Larry Gallagher, "Stone Age Beer," *Discover,* November 2005.

9. William Shurtleff and Akiko Aoyagi, *History of Kogi—Grains and/or Soybeans Enrobed with a Mold Culture (300 BCE to 1212)* (Lafayette, Calif.: Soyinfo Center, 2012).

10. Michal Pollan, *The Botany of Desire* (New York: Random House, 2002, Kindle edition).

11. For a discussion of the relative health of foragers and farmers see Mark Nathan Cohen, *Health and the Rise of Civilization* (New Haven: Yale University Press, 1980).

12. See, e.g., Robert J. Braidwood, Jonathan D. Sauer, Hans Helbaek, et al., "Did Man Once Live by Beer Alone?" *American Anthropologist* n.s. 55, no. 4 (October 1953): 515–26.

13. Statistics Division, Agriculture Organization of the United Nations, http://www.fao.org/statistics/en/ (accessed 24 October 2014).

14. "Enforcement Report," U.S. Food and Drug Administration, 29 August 2001, available at http://web.archive.org/web/20070613081904/http://www.fda.gov/bbs/topics/ENFORCE/2001/ENF00708.html.

15. "The Global Wine Industry, Slowly Moving from Balance to Shortage," report from Morgan Stanley, 22 October 2013, available at http://blogs.reuters.com/counterparties/files/2013/10/Global-Wine-Shortage.pdf

16. Tunde Jurikova, Jiri Sochor, Otakar Rop, et al., "Polyphenolic Profile and Biological Activity of Chinese Hawthorn," *Molecules* 17, no. 12 (2012): 14490–15509.

17. Denham Harman, "Aging: A Theory Based on Free Radical and Radiation Chemistry," *Journal of Gerontology* 11 (1956): 298–300.

18. Ruth H. Matthews, Pamel R. Pehrsson, and Mojgan Forhat-Sabet, "Sugar Content of Selected Foods," U.S. Department of Agriculture, Home Economics Research Report, no. 48 (September 1987); He Guifen, Jialin Sui, Jinhua Du, and Jing Lin, "Characteristics and Antioxidant Capacities of Five Hawthorn Wines Fermented by Different Wine Yeasts," *Journal of the Institute of Brewing* 119, no. 4 (October 2013): 321–27.

19. Simon Singh and Edzard Ernst, *Trick or Treatment? Alternative Medicine on Trial* (London: Transworld, 2009, Kindle edition).

20. Henry C. Lu, *Traditional Chinese Medicine: An Authoritative and Comprehensive Guide* (Laguna Beach, Calif.: Basic Health Publications, 2005), 13.

21. Subhuti Dharmananda, "Hawthorn: Food and Medicine in China" (Institute for Traditional Medicine, Portland, Ore.), http://www.itmonline.org/arts/crataegus.htm.

22. Li Zhisuie, *The Private Life of Chairman Mao* (New York: Random House, 2011), 542.

23. James Reston, "Now Let Me Tell You About My Appendectomy in Peking," *New York Times,* 26 July 1972, 1.

24. T. V. N. Persuad, *Early History of Anatomy: From Antiquity to the Beginning of the Modern Era* (Springfield, Ill.: Thomas, 1984), 20–22.

25. World Health Organization, "Acupuncture: Review and Analysis of Reports on Controlled Clinical Trials" (2003). For a summary of the criticism of the WHO report see Singh and Ernst, *Trick or Treatment?*

26. Weillang Weng, W. Q. Zhang, F. Z. Liu, et al., "Therapeutic Effect of *Crataegus pinnatifida* on 46 Cases of Angina Pectoris. A Double Blind Study," *Journal of Traditional Chinese Medicine* 4, no. 4 (1984): 293–94.

EIGHT
The Medicine Tree

1. Heart disease statistics are taken from the following sources: World Health Organization, "Fact Sheet No. 310," July 2013; Centers for Disease Control and Prevention, Atlanta, Georgia; "Fast Stats," American Heart Association, *Circulation* (2011); European Society of Cardiology, the European Heart Network, and the British Heart Foundation Health Promotion Research Group, *European Cardiovascular Disease Statistics,* 4th ed. (Department of Public Health, University of Oxford).

2. Pedanius Dioscorides, *De medica materia,* trans. Tess Anne Osbaldeston, books 1 and 2; P. De Vos, "European Materia Medica in Historical Texts: Longevity of a Tradition and Implications for Future Use," *Journal of Ethnopharmacology* 132, no. 1 (October 2010): 28–47.

3. David Pybus and Charles Sell, eds., *The Chemistry of Fragrances* (London: Royal Society of Chemistry, 1999), 15.

4. Stephen Harrod Buhner, *Sacred Plant Medicine* (Rochester, Vt.: Inner Traditions, 2006), 128.

5. "*Crataegus oxycantha* in the Treatment of Heart Disease," letter from Dr. M. C. Jennings, *New York Medical Journal,* 10 October 1896.

6. Joseph Clements, "*Crataegus oxycantha* in Angina Pectoris with Report of a Case," *Kansas City Medical Index* 19, no. 5 (1898): 131–32.

7. Material about the Eclectics, Lloyd Brothers, and John Uri Lloyd is from John S. Haller, *Medical Protestants: The Eclectics in American Medicine, 1824–1939* (Carbondale: Southern Illinois University Press, 2013), and Michael A. Flannery, *John Uri Lloyd: The Great American Eclectic* (Carbondale: Southern Illinois University Press, 1998).

8. Material about the therapeutic use of *Crataegus* is from editorials by Finley Ellingwood in his journal *Ellingwood's Therapeutist* 8 (1914): 71, and 9 (1915): 217–

18; John Uri Lloyd, "A Treatise on *Crataegus,*" Drug Treatise 16 (Cincinnati: Lloyd Brothers, 1917), and Harvey Wilkes Felter, *The Eclectic Materia Medica, Pharmacology and Therapeutics* (Repr.; Sandy, Ore.: Eclectic Medical Publications, 1983), 130.

9. Finley Ellingwood, ed., *Ellingwood's Therapeutist* 11 (1917): 21.

10. "WHO Model List of Essential Medicines," World Health Organization, October 2013. John Uri Lloyd also wrote novels set in northern Kentucky, where he grew up. The most famous of these was an illustrated work of science fiction called *Etidorpha,* which was widely popular in Europe and America. Part of the genre of "hollow earth" novels such as Jules Verne's *Journey to the Center of the Earth,* the books spawned literary clubs all over the United States. Parents even named their daughters Etidorpha (An anagram of "Aphrodite"). It has been speculated that Lloyd's imagery—huge subterranean lakes, chambers full of giant mushrooms, and wise, pale beings who live underground—was influenced by his knowledge of psychoactive substances such as marijuana, psilocybin and opium; see Marcus Boon, *The Road of Excess: A History of Writers on Drugs* (Cambridge: Harvard University Press, 2002), 228.

11. See Haller, *Medical Protestants;* lists of schools are available at http://www.lcme.org/directory.htmandhttp://www.osteopathic.org/inside-aoa/about/affiliates/Pages/osteopathic-medical-schools.aspx (osteopathic colleges); Ron Paul, speech before the U.S. House of Representatives, *Congressional Record* 155, part 17, 23 September 2009–6 October 2009.

12. James D. P. Graham, "*Crataegus oxycantha* in Hypertension," *British Medical Journal,* 11 November1939, 951.

13. For material about mesmerism and Benjamin Franklin's role in the Paris experiment see Mark A. Best, *Benjamin Franklin: Verification and Validation of the Scientific Process in Healthcare as Demonstrated by the Report of the Royal Commission of Animal Magnetism* (Bloomington, Ind.: Trafford, 2003), Robert Darnton, *Mesmerism and the End of Enlightenment in France* (Cambridge: Harvard University Press), and Russell Shorto, *Descartes' Bones: A Skeletal History of the Conflict Between Faith and Reason* (New York: Random House, 2008).

14. Best, *Benjamin Franklin,* 12.

15. E-mail to the author from Dennis Schlagheck, Morton College Library and Hawthorne Works Museum of Morton College, Cicero, Ill., 24 January 2014.

16. Jonathan Freedman and David O. Sears, *Social Psychology,* 4th ed. (Englewood Cliffs, N.J.: Prentice-Hall, 1981); Steven D. Levitt and John A. List, "Was There Really a Hawthorne Effect at the Hawthorne Plant? An Analysis of the Original Illumination Experiments," *American Economics Journal* (January 2011): 224–38.

17. Charles Sanders Peirce and Joseph Jastrow, "On Small Differences in Sensation," *Memoirs of the National Academy of Sciences* 3 (1885): 73–83.

18. "Streptomycin Treatment of Pulmonary Tuberculosis," *British Medical Journal* (30 October 1948): 769–82; J. L. Vanneste, ed., *Fire Blight: The Disease and Its Causative Agent,* Erwinia amylovora (Wallingford, U.K.: CABI, 2000), 22. Streptomycin has been used in the United States to combat fire blight, a serious

bacterial disease of hawthorns, pears, apples, and other members of the rose family, although, like tuberculosis, some strains of this bacteria have evolved an alarming resistance to the antibiotic.

19. G. Zapfe, "Clinical Efficacy of *Crataegus* Extract WS 1442 in Congestive Heart Failure NYHA Class II," *Phytomedicine* 8 (2001): 262–66. For an analysis of the clinical trials summarized in this section see an overview of research compiled by Mary C. Tassell, , Rosari Kingston, Ambrose Furey, et al., "Hawthorn (*Crataegus*) in the Treatment of Cardiovascular Disease," *Pharmacognosy Review* (January–June 2010): 32–41.

20. Monograph on *Crataegus* released by Kommission E, 19 July 1994, available at http://buecher.heilpflanzen-welt.de/BGA-Kommission-E-Monographien/crataegi-flos-weissdornblueten.htm.

21. C. J. Holubarsch et al., "The Efficacy and Safety of Crataegus Extract WS 1442 in Patients with Heart Failure: The SPICE Trial," *European Journal of Heart Failure* 12 (10 December 2008): 1255–63.

22. M. Tauchert, "Efficacy and Safety of *Crataegus* Extract WS 1442 in Comparison with Placebo in Patients with Chronic Stable New York Heart Association Class-III Heart Failure," *American Heart Journal* 143, no. 5 (2002): 910–15.

23. Mary C. Tassell et al., "Hawthorn (*Crataegus spp.*) in the Treatment of Cardiovascular Disease," *Pharmacognosy Review* 4, no. 7 (January–June 2010): 32–41.

24. Ibid.; Dr. Keith Aaronson, telephone conversation with the author, 17 October 2005.

25. "University Suspects Fraud by a Researcher Who Studied Red Wine," *New York Times,* 11 January 2012, A15; Malcolm Law and Nicholas Ward, "Why Heart Disease Mortality Is Low in France: The Time Lag Explanation," *British Medical Journal* 318 (29 May 1999): 1471–80.

26. Eliseo Guallar, Saverio Stranges, Cynthia Mulrow, et al., "Enough Is Enough: Stop Wasting Money on Vitamin and Mineral Supplements," *Annals of Internal Medicine* 159, no. 12 (17 December 2013): 850–51; "Resveratrol May Not Be the Elixir in Red Wine and Chocolate," May 13, 2014, *NPR,* http://www.npr.org/blogs/thesalt/2014/05/13/311904587/resveratrol-may-not-be-the-elixir-in-red-wine-and-chocolate.

27. Statistics compiled by Symphony IRI Group (Chicago, 2013), available at the American Botanical Council Web site, http://cms.herbalgram.org/herbalgram/issue99/hg99-mktrpt.html.

28. Josh Long, "FDA GMP Inspectors Cite 70% of Dietary Supplement Firms," *Natural Products Insider,* 20 May 2013 (online journal).

29. J. Si, G. Gao, and D. Chen, "Chemical Constituents of the Leaves of *Crataegus scabrifolia,*" *Zhongguo Zhong Yao Za Zhi,* 23 July 1998, 448.

30. Haller, *Medical Protestants;* e-mail to the author from Kristin Kile of the Eclectic Institute, 26 December 2013; Rex Sallabanks, "Fruiting Plant Attractiveness to Avian Seed Dispersers: Native vs. Invasive *Crataegus* in Western Oregon," *Madroño* 40, no. 2 (April–June 1993): 108–16.

31. For a description of the trial see M. Hanus, J. Lafon, and M. Mathieu,

"Double-blind, Randomised, Placebo-controlled Study to Evaluate the Efficacy and Safety of a Fixed Combination Containing Two Plant Extracts (*Crataegus oxyacantha* and *Eschscholtzia californica*) and Magnesium in Mild-to-moderate Anxiety Disorders," *Current Medical Research and Opinion* 20 (January 2004): 63–71. For a criticism of the trial see Sy Atezaz Saeed, Richard M. Bloch, and Diana J. Antonacci, "Herbal and Dietary Supplements for Treatment of Anxiety Disorders," *American Family Physician* 76, no. 4 (August 2007): 549–56.

32. Paracelsus, *Liber Paragranu,* 1531.

33. Letter to the editor, "My horse is eating hawthorn is this bad for him?" *Horse and Rider,* n.d., at http://www.horseandrideruk.com/article.php?id=889.

34. Ruoling Guo, Max H. Pittler, and Edzard Ernst, "Hawthorn Extract for Treating Chronic Heart Failure," Cochrane Database of Systematic Reviews, 23 January 2008.

35. University of Maryland Medical Center, "Hawthorn," online report, http://www.umm.edu/health/medical/altmed/herb/hawthorn (accessed 29 October 2014).

36. David Healy, *Pharmageddon* (Berkeley: University of California Press, 2012), 5–6; "GlaxoSmithKline to Stop Paying Doctors to Promote Drugs," *The Guardian,* 17 December 2013, Business Section.

37. Rhonda M. Cooper-DeHoff, Yan Gong, Eileen M. Handberg, et al., "Tight Blood Pressure Control and Cardiovascular Outcomes Among Hypertensive Patients with Diabetes and Coronary Artery Disease," *Journal of the American Medical Association* 304, no. 1 (July 2010): 61–68.

NINE
A Tree for All Seasons

1. A. Barnea and F. Nottebohm, "Seasonal Recruitment of Hippocampal Neurons in Adult Free-ranging Black-capped Chickadees," *Proceedings of the National Academy of Sciences of the United States of America* 91, no. 23 (November 1994): 11217–21.

2. Manji S. Dhindsa and David A. Boag, "Patterns of Nest Site, Territory, and Mate Switching in Black-billed Magpies (*Pica pica*)," *Canadian Journal of Zoology* 70, no. 4 (1992): 633–40.

3. For more about magpies see Gisela Kaplan, "Song Structure and Function of Mimicry in the Australian Magpie (*Gymnorhina tibicen*) Compared to the Lyrebird (*Menura ssp.*)" *International Journal of Comparative Psychology* 12, no. 4 (1999): 219–41, and "Black-billed Magpie," Toronto Zoo Web site, www.torontozoo.com/exploretheZoo/AnimalDetails.asp?pg=546.

4. Gary J. Wiles, "Records of Anting by Birds in Washington and Oregon," *Washington Birds* 11 (2011): 28–34.

5. Helmut Prior, Ariane Schwarz, Onur Güntürkün, et al., "Mirror-Induced Behavior in the Magpie (*Pica pica*): Evidence of Self-Recognition," *PLOS Biology* (19 August 2008). Although the magpies at Dark Acres belong to the species *Pica hudsonia,* Helmut Prior told the author in a 19 May 2013 e-mail that the genetic

differences between our black-billed magpies and the Eurasian magpies he studied are slight, and their behavioral ecology "seems to be more or less the same."

6. Marc Bekoff, *The Emotional Lives of Animals: A Leading Scientist Explores Animal Joy, Sorrow, and Empathy—and Why They Matter* (Novato, Calif.: New World Library, 2009), 1–2.

7. Davorin Tome, "Changes in the Diet of Long-Eared Owl *Asio otus:* Seasonal Patterns of Dependence on Vole Abundance," *Ardeola* 56, no. 1 (2009): 49–56.

8. Sue Manning, "Infrared Camera in Wild Aimed at Montana Owl Nest," Associated Press, 11 April 2013.

9. Holt can be seen in the video posted online at http://www.owlinstitute.org/owl-cam.html.

10. Norbert Lefranc, *Shrikes* (London: A&C Black, 2013).

11. Piotr Tryjanowski and Martin Hromada, "Do Males of the Great Grey Shrike, *Lanius excubitor,* Trade Food for Extrapair Copulations?" *Animal Behavior* 69, no. 3 (March 2005): 529–33.

12. Jane Rider, "Northern Shrike Catches Vermillion Flycatcher," *Missoulian,* online edition, 13 January 2000.

13. For more about the chemical warfare of plants see Elroy L. Rice, *Allelopathy* (New York: Academic Press, 1984), and Zahid A. Cheema, *Allelopathy* (New York: Springer, 2012).

14. Roger del Moral and Cornelius H. Muller, "The Allelopathic Effects of *Eucalyptus camaldulensis,*" *American Midland Naturalist* 83, no. 1 (January 1970): 254–82.

15. Rob Chaney, "City Sees Some Success Removing Norway Maples from Greenough Park," *Missoulian,* 28 September 2011.

16. Ibid.

17. For more about cedar-hawthorn rust see John R. Hartman, Thomas P. Pirone, and Mary Ann Sall, *Pirone's Tree Maintenance* (Oxford: Oxford University Press, 2000), 273.

18. J. L. Vanneste, ed., *Fire Blight: The Disease and Its Causative Agent,* Erwinia amylovora (Wallingford, U.K.: CABI, 2000).

19. For more about this process of evolution see J. L. Feder, "The Apple Maggot Fly, *Rhagoletis pomonella:* Flies in the Face of Conventional Wisdom About Speciation?" in *Endless Forms: Species and Speciation,* ed. D. J. Howard and S. H. Berlocher (Oxford: Oxford University Press, 1998).

20. Roger L. Blackman, *Aphids* (Cambridge: Ginn, 1974).

21. L. Hugh Newman, *Living with Butterflies* (London: Baker, 1967), 208. To see a video about the Camberwell Beauty release go to http://www.britishpathe.com/video/camberwell-butterfly.

22. Kenneth A. Pivnick and Jeremy N. McNeil, "Puddling in Butterflies: Sodium Affects Reproductive Success in *Thymelicus lineola,*" *Physiological Entomology* 12, no. 4 (December 1987): 461–72.

23. "Restoration of Douglas Hawthorn (*Crataegus douglasii*) in the Great Lakes Region," USDA Forest Service Web site, www.fs.fed.us/wildflowers/Rare_

Plants/conservation/success/crataegus_douglasii_restoration.shtml (accessed 27 October 2014).

TEN

Essence and Spinescence

1. Alonso Ricardo and Jack W. Szostak, "The Origin of Life on Earth," *Scientific American,* September 2009, 54–61.

2. Paul Blum, *Archaea: Ancient Microbes, Extreme Environments, and the Origin of Life* (New York: Gulf Professional, 2001).

3. For more about the evolution of life on earth and the first plants and trees see Colin Tudge, *The Tree* (New York: Crown, 2005), and Wilson N. Stewart and Gar W. Rothwell, *Paleobotany and the Evolution of Plants* (Cambridge: Cambridge University Press, 1993).

4. C. Kevin Boyce, Carol L. Hotton, Marilyn L. Fogel, et al., "Devonian Landscape Heterogeneity Recorded by a Giant Fungus," *Geology* 35 (May 2007): 399–402.

5. See Craig L. Schmitt and Michael L. Tatum, *The Malheur National Forest, Location of the World's Largest Living Organism [The Humongous Fungus]* (N.p.: U.S. Department of Agriculture, Forest Service, 2008).

6. William E. Stein, Frank Mannolini, Linda VanAller Hernick, et al., "Giant Cladoxylopsid Trees Resolve the Enigma of the Earth's Earliest Forest Stumps at Gilboa," *Nature,* 19 April 2007, 904–7.

7. Michael C. Wiemann and David W. Green, "Estimating Hardness from Specific Gravity for Tropical and Temperate Species," Research Paper FPL-RP-643, U.S. Department of Agriculture, September 2007.

8. Michael Wiemann, telephone conversation with the author, 12 November 2013.

9. Ohio Secretary of State, "Annual Report" (1887), 579–80.

10. On "witches brew" see W. H. Camp, "The *Crataegus* Problem," Castanea 7, nos. 4–5 (April–May 1942). On "taxonomic disaster" see S. B. Sutton, *Charles Sprague Sargent and the Arnold Arboretum* (Cambridge: Harvard University Press, 1970).

11. James Bird Phipps, "*Mespilus canescens,* a New Rosaceous Endemic from Arkansas," *Systematic Botany* 15, no. 1 (January–March 1990): 26–32.

12. Gerard Krewer and Tom Crocker, "Experiments and Observations on Growing Mayhaws as a Crop in South Georgia and North Florida," Cooperative Extension Service, the University of Georgia College of Agricultural and Environmental Sciences, 2009.

13. For a succinct explanation of the tejocote problem see the David Karp, "Tejocote: No Longer Forbidden," *Fruit Gardener* 42, no. 6 (November and December 2010): 10–13.

14. Luther Burbank et al., "Luther Burbank: His Methods and Discoveries and Their Practical Application," vol. 6 (New York: Luther Burbank Press, 1914), 266.

15. Ibid.

16. See Karp, "Tejocote: No Longer Forbidden,"; E-mail from the USDA in response to the author's question about the status of tejocote imports from Mexico, 9 December 2013.

17. *Creating Deciduous Bonsai Trees—Hawthorn,* video available at http://www.imeo.com/36470902.

18. Graham Potter e-mail to the author, 28 November 2013.

19. R. J. Rayner, "New finds of *Drepanophycus spinaeformis Göppert* from the Lower Devonian of Scotland," *Transactions of the Royal Society of Edinburgh: Earth Sciences* 75, no. 4 (January 1984): 353–63.

20. Peter Marks, "Reading the Landscape," *Cornell Plantations Magazine* (summer 2001); Helen M. Armstrong, Liz Poulsom, Tom Connolly, and Andrew Peace, "A Survey of Cattle-grazed Woodlands in Britain," Forest Research, Northern Research Station, Roslin, Midlothian, Scotland, October 2003.

21. Colin Beale, "Why Grow Thorns if They Don't Work?" *Nothing in Biology Makes Sense,* http://nothinginbiology.org/2012/02/09/why-grow-thorns/.

22. Colin Beale, e-mail to the author, 5 November 2013.

23. Malka Halpern, Dina Raats, and Simcha Lev-Yadum, "The Potential Anti-Herbivory Role of Microorganisms on Plant Thorns," *Plant Signaling and Behavior* 2, no. 6 (November–December 2007), 503–4.

24. T. P. Olenginski, "Plant Thorn Synovitis: An Uncommon Cause of Monoarthritis," *Seminars in Arthritis and Rheumatism* 1 (21 August 1991): 40–46.

25. John Cartwright, *Evolution and Human Behavior: Darwinian Perspectives on Human Nature* (Cambridge: MIT Press, 2000), 97–98; Katie Alcock, "Ants Work with Acacia Trees to Prevent Elephant Damage," *BBC News,* 2 September 2010.

26. Dan O'Neill, *The Last Giant of Beringia: The Mystery of the Bering Land Bridge* (New York: Basic, 2009).

27. James Bird Phipps, "Biogeographical, Taxonomic and Cladistic Relationships Between Eastern Asiatic and North American *Crataegus,*" *Annals of the Missouri Botanical Garden* 70 (1983): 666–700.

28. For the theory that the Americas were settled by people traveling south from Beringia by foot see O'Neill, *Last Giant of Beringia;* for the theory that they traveled by sea see James E. Dixon, *Quest for the Origins of the First Americans* (Albuquerque: University of New Mexico Press, 1991).

ELEVEN
The Crown of Thorns

1. Richard Savil, "Vandals Destroy Holy Thorn Tree in Glastonbury," *The Telegraph,* 9 December 2010.

2. For the myth of Joseph, see Rabanus Maurus, *Life of Mary Magdalene,* a ninth century manuscript discovered a thousand years later and published in 1842; Robert de Boron, *Joseph d'Arimathie* (c. 1350); and Valeria M. Lagorio, "The Evolving Legend of St Joseph of Glastonbury," *Speculum* 46, no. 2 (April 1971): 209.

3. Giraldus Cambrensis (Gerald of Wales), *Liber de Principis Instructione* (c. 1193); Geoffrey Ashe, *King Arthur's Avalon* (Gloucestershire, U.K.: History Press, 2007).

4. *Glastonbury: Isle of Light,* Galatia Film Company, Web site, http://www.glastonburyfilm.com/; "Liam Neeson Could Star in Glastonbury Movie Shot in Kazakhstan," *Western Daily Press,* 7 September 2012.

5. Rodney Castleden, *King Arthur: The Truth Behind the Legend* (London: Routledge, 2003); Francis Aidan Gasquet, *Last Abbot of Glastonbury and Other Essays* (London: George Bell and Sons, 1908), 68–71.

6. K. I. Christensen, "Revision of *Crataegus,*" in the section "*Crataegus* in the Old World," *Systematic Botany Monographs* 35 (1992): 1–199; James Bird Phipps, *Hawthorns and Medlars* (Portland, Ore.: Timber Press, 2003), 16.

7. See Savil, "Vandals Destroy Holy Thorn Tree in Glastonbury."

8. Luke Salked, "Were Anti-Christians Behind Pilgrimage Site Attack? 2,000-year-old Holy Thorn Tree of Glastonbury Is Cut Down," *Mail Online,* 9 December 2010; "Killed Off After 2,000 Years: Glastonbury's Vandalized Holy Thorn Tree Must Be Replaced After 'Trophy Hunters' Snap Off Its New Shoots, *Mail Online,* 19 September 2011; "The Story of the Glastonbury Thorn: The Recent Chapters," Glastonbury Pilgrim Reception Centre Web site: http://www.unitythroughdiversity.org/the-recent-chapters.html.

9. Geoffrey Humphreys, *History Today* 48, no. 12 (1998); Web site of the Greenmantle Nursery, Garberville, Calif., http://www. greenmantlenursery.com; Peter Marteka, "Town Hoping a Legend Takes Root but Outlook Isn't Rosy for Glastonbury Thorn," *Hartford Courant,* 14 April 2006; Adrian Hamilton, "Will the Last Person to Leave the Church of England Please Turn Off the Lights," *Independent,* 18 April 2011.

10. Ronald Hutton, *The Stations of the Sun: A History of the Ritual Year in Britain* (Oxford: Oxford University Press, 1996), 218–25; Kathryn Price NicDhàna, Erynn Rowan Laurie, C. Lee Vermeers, and Kym Lambert ní Dhorreann, *The CR FAQ: An Introduction to Celtic Reconstructionist Paganism* (Memphis, Tenn.: River House, 2007), 100–3.

11. Paul Barber, *Vampires, Burial, and Death: Folklore and Reality* (New Haven: Yale University Press, 1988), 48 and 72; Friedrich S. Krauss, "South Slavic Countermeasures Against Vampires," in *The Vampire: A Casebook,* ed. Alan Dundes (Madison: University of Wisconsin Press, 1998).

12. John Mandeville, *The Voyages and Travels of Sir John Mandevile, Knight* (printed by A. Wilde, for G. Conyers, in Little-Britain, T. Norris, at London-bridge, and A. Bettesworth, in Pater-Noster-Row, 1722), 9–10; John Larner, "Plucking Hairs from the Great Cham's Beard: Marco Polo, Jan de Langhe, and Sir John Mandeville," in *Marco Polo and the Encounter of East and West,* ed. Suzanne Conklin Akbari and Amilcare Iannucci (Toronto: University of Toronto Press, 2008), 133–55.

13. *All the Year Round,* 20 April 1889, 371.

14. Nerman Maharik, Elgengaihi Souad, and Tahaa Hussein, "In Vitro Mass

Propagation of the Endangered Sinai Hawthorn, *Crataegus sinaica boiss,*" *International Journal of Academic Research* 1, no. 1 (January 2009): 24.

15. *The Catholic Encyclopedia: An International Work of Reference on the Constitution, Doctrine, Discipline, and History of the Catholic Church,* vol. 4 (New York: Appleton, 1913), 540–41; Michael F. Hendy, *Studies in the Byzantine Monetary Economy, c. 300–1450* (Cambridge: Cambridge University Press, 2008), 230; Donald M. Nicol, *Byzantium and Venice: A Study in Diplomatic and Cultural Relations* (Cambridge: Cambridge University Press, 1992).

16. "Thorn from Jesus's Crucifixion Crown Goes on Display at British Museum," *Mail Online,* 24 March 2011.

17. Joan Carroll Cruz, *Relics* (Huntington, Ind.: One Sunday Visitor, 1984), 36–37; Avidoam Danin, Alan D. Whanger, and Mary Whanger, *Flora of the Shroud of Turin* (St. Louis: Missouri Botanical Garden, 1999); Mary Whanger and Alan D. Whanger, *The Shroud of Turin: An Adventure of Discovery* (Franklin, Tenn.: Providence House, 1998). Another popular biblical mystery concerns the burning bush that appeared to Moses. According to the account in Exodus it burned yet wasn't consumed by the flames, and from the inferno came first the voice of an angel of God and then of God himself, telling Moses to take off his sandals because he was standing on holy ground. Moses hid his face, afraid to look at God. The Hebrew word used to describe the bush is usually translated "brambles." There is some speculation that this account was based on experiences with a lemony-smelling flowering plant called white dittany or burning bush, *Dictamnus albus*. In the summer it secretes a flammable oil that can ignite with a flash but causes no harm to the flowers. In medieval and Renaissance Europe, however, it was widely believed that what Moses averted his eyes from was a burning hawthorn. (On dittany, see Missouri Botanical Garden Web site, http://www.missouribotanicalgarden.org/PlantFinder/PlantFinderDetails.aspx?kempercode=c490; on the attribution of the hawthorn as the burning bush see "Myths and Legends of Hawthorn Trees" at http://www.paghat.com/hawthornmyths.html [accessed 2 November 2014].)

18. Francois Leuret, *Modern Miraculous Cures: A Documented Account of Miracles and Medicine in the Twentieth Century* (Bangalore: Hesperides, 2008), 63.

19. Rosemary Guiley, *The Encyclopedia of Saints* (New York: Infobase, 2001), 389.

20. "Our Lady of Beauraing," http://www.marypages.com/beauraingEng1.htm; "What Happened to the Children?" Web site about the Beauraing children as adults, http://beauraing.catho.be/uk/uk_430_enfants.html.

21. In some ways similar to rag trees among Northern Plains Indians is the Medicine Wheel in northern Wyoming, a circle of stones eighty feet in diameter with twenty-eight stone spokes radiating from the circumference to the center. Fixed to the rope fence erected to mark its perimeter are beads, pouches, braids of sweetgrass, and mysterious bundles wrapped in rawhide. While it is not known how old it is or who built it, the Medicine Wheel, sitting ten thousand feet up a windy slope in the Bighorn Mountains, is a sacred site central to the traditional spirituality of the tribes of the Northern Plains. An Arikira elder named Old Mouse

explained, "Eventually one gets to the Medicine Wheel to fulfill one's life." See Andrew Gulliford, *Sacred Objects and Sacred Places: Preserving Tribal Traditions* (Boulder: University Press of Colorado, 2000), 135–44.

22. "Best Place to Make a Wish," *The Scotsman*, 17 October 2007, available at http://www.scotsman.com/lifestyle/outdoors/best-place-to-make-a-wish-1-696724.

23. Virginia Yans-McLaughlin, *Immigration Reconsidered: History, Sociology, and Politics* (Oxford: Oxford University Press, 1990), 100.

24. Christine Zucchelli, *Trees of Inspiration: Sacred Trees and Bushes of Ireland* (Wilton, Ireland: Collins, 2009).

25. Gordon Deegan, "Clare Fairy Tree Vandalized," *Irish Times*, 14 August 2002.

26. Maeve Binchy, *Whitethorn Woods* (New York: Knopf, 2007).

27. Nick Sutton, *The DeLorean Story: The Car, the People, the Scandal* (Sparkford, U.K.: Haynes, 2013), 69.

TWELVE
The Warrior Queen

1. American Forests Web site, http://www.americanforests.org/our-programs/global-releaf-projects/.

2. To learn more about the life, work, and mysterious death of David Douglas see M. L. Tyrwhitt-Drake, "Douglas, David," in *Dictionary of Canadian Biography*, vol. 6 (Toronto: University of Toronto, 1987); David Douglas, *Journal Kept by David Douglas During his Travels in North America, 1823–1827* (London: William Wesley, 1914); Jean Greenwell, "Kaluakauka Revisited: The Death of David Douglas in Hawai'i," *Hawaiian Journal of History* 22 (1988): 147–69.

3. James Bird Phipps, *Hawthorns and Medlars* (Portland, Ore.: Timber Press, 2003).

4. Much of the material about Phipps came by way of e-mails between Phipps and the author in 2014.

5. W. H. Camp, "The *Crataegus* Problem," *Castanea* 7, nos. 4–5 (April–May, 1942): 51–55.

6. Eugenia Y. Y. Lo, Saša Stefanović, and Timothy A. Dickinson, "Molecular Reappraisal of Relationships Between *Crataegus* and *Mespilus* (*Rosaceae, Pyreae*)—Two Genera or One?" *Systematic Botany* 32, no. 3 (2007): 596–616.

7. Alex V. Popovkin, Katherine G. Mathews, José Carlos Mendes Santos, et al., "*Spigelia genuflexa* (Loganiaceae), a New Geocarpic Species from the Atlantic Forest of Northeastern Bahia, Brazil," *PhytoKeys* 6 (2011): 47–56.

Epilogue

1. Peter Berresford Ellis, *A History of the Irish Working Class* (London: Pluto), 123–24.

2. Sue Moran, "Thomas Moran" (2004), unpublished manuscript; copy in the author's possession.

3. Robert Vaughn, *Then and Now: Thirty-six Years in the Rockies* (Helena, Mont.: Far Country Press, 2001), 97.

4. Genevieve McBride, *The Bird Tail* (New York: Vantage, 1974).

5. W. C. Jameson, *Buried Treasures of the Rocky Mountain West: Legends of Lost Mines, Train Robbery Gold, Caves of Forgotten Riches and Indians' Buried Silver* (Atlanta: August House, 1993), 87–93.

Acknowledgments

I thank Michelle Tessler for her expert crafting of the proposal that led to this book. For their help and hospitality I'm also grateful to Jane and James Greenfield in Pennsylvania, Willy Hogan in Ireland, Jim Cornelius at St. Peter's in Montana, Mary and Darvin Dent in Kansas, Marcia and Victor Lieberman in California, and Marcia Herrin and Alan Strickland in New Hampshire. By sharing his experiences of researching a complicated book, Bryan Di Salvatore spared me the frustration of false starts and dead ends. Finally, I'm indebted to Maeve for apparently understanding the intention of the things I hung from her branches: the computer keyboard, the scraps of cloth, and especially Kitty's necklace.

Index